ABUNDANCE EFFECTS IN CLASSIFICATION

W. W. MORGAN

INTERNATIONAL ASTRONOMICAL UNION

UNION ASTRONOMIQUE INTERNATIONALE

SYMPOSIUM No. 72

HELD IN LAUSANNE-DORIGNY, SWITZERLAND, JULY 8-11, 1975

ABUNDANCE EFFECTS
IN CLASSIFICATION

EDITED BY

B. HAUCK

*Institut d'astronomie de l'Université de Lausanne
et Observatoire de Genève, Switzerland*

AND

P. C. KEENAN

Perkins Observatory, Delaware, U.S.A.

D. REIDEL PUBLISHING COMPANY

DORDRECHT-HOLLAND / BOSTON-U.S.A.

1976

Cloth edition: ISBN 90-277-0674-3
Paperback edition: ISBN 90-277-0675-1

Published on behalf of
the International Astronomical Union
by
D. Reidel Publishing Company, P.O. Box 17, Dordrecht, Holland

Sold and distributed in the U.S.A., Canada, and Mexico
by D. Reidel Publishing Company, Inc.
Lincoln Building, 160 Old Derby Street, Hingham,
Mass. 02043, U.S.A.

Printed in The Netherlands

Dedicated to W. W. MORGAN

TABLE OF CONTENTS

PREFACE   XI

THE ORGANIZING COMMITTEES   XII

LIST OF PARTICIPANTS   XIII

WELCOME TO W. W. MORGAN *by B. Strömgren*   XV

BIBLIOGRAPHY OF W. W. MORGAN   XIX

PART I / INFLUENCE OF ABUNDANCES UPON STELLAR ATMOSPHERE CALCULATIONS

B. BASCHEK *(Invited Lecture)*   3

A. W. IRWIN, C. T. BOLTON, and R. F. GARRISON / Synthesis of Classification Dispersion Spectra from Stellar Atmospheres Theory and from High Dispersion Spectra   17

R. MÄCKLE, H. HOLWEGER, and R. and R. GRIFFIN / A Model-Atmosphere Analysis of the Spectrum of Arcturus   19

A. J. SAUVAL / Influence of Uncertainties of Molecular Data upon the Determination of Abundances in Cool Stars   21

I. BUES / On the Spectral Classification of Cool White Dwarfs   23

R. WEHRSE / Effects of Abundance Changes on the Spectra of Very Cool White Dwarfs   25

PART II / SOME COMMENTS ON A CATALOGUE OF ATMOSPHERIC PARAMETERS AND [Fe/H] DETERMINATIONS

G. CAYREL DE STROBEL *(Invited Lecture)*   29

PART III / DERIVATION OF ABUNDANCES THROUGH PHOTOMETRIC AND SPECTROSCOPIC METHODS

R. A. BELL *(Invited Lecture)*   49

B. HAUCK / Abundance Effects for the A0-G2 Stars in the Geneva System   67

D. L. CRAWFORD and C. L. PERRY / Abundance Effects on *uvby* Photometry   71

J. J. CLARIÁ and W. OSBORN / A Test of the Accuracy of Low Dispersion
Objective Prism: Spectral Classification of Late-Type Stars Using DDO Photometry   73

M. GRENON / Observed Relations between Physical Properties and MK Classification as Functions of Metal Abundance for F5-K4 Stars   75

E. E. MENDOZA V. / Narrow-Band Photometry of A and F Stars   79

P. E. NISSEN / The Helium-to-Hydrogen Ratio of Stars in Young Clusters and
Associations   81

N. R. WALBORN / Recent Results Pertaining to the Helium-Rich Stars   85

P. M. WILLIAMS / Abundance Effects in the Classification of K Stars   87

C. PAYNE-GAPOSCHKIN / Comparison of Colour Curves of Mira Stars of Spectrum M and S   91

R. CANAVAGGIA, P. MIANES, and J. ROUSSEAU / The Colour Excess Scale
and Intrinsic Colour Properties of the Longer Period Cepheids   95

A. MAEDER / On the Derivation of Abundances by means of Features in the
Sequences of the Old Open Star Clusters   97

W. OSBORN and J. J. CLARIÁ / A Method for Determining the Chemical Composition Parameter (X, Y, Z) of Galactic Clusters   101

PART IV / ABUNDANCE EFFECTS IN SPECTRAL CLASSIFICATION

C. JASCHEK *(Invited Lecture)*   113

N. HOUK and M. R. HARTOOG / HD Stars South of $\delta = -53°$ Having Peculiar
Abundances: Statistics and Notation   127

W. C. SEITTER / The Bonner Spectral Atlas and Three-Dimensional Classification
at Low Dispersions   129

A. PRZYBYLSKI / The Spectral Type of HD 101065   135

D. CHALONGE and L. DIVAN / Classification of Stars with Different Chemical   143
Constitution

P. C. KEENAN / Effects of Heavy-Element Abundance on Classification of G-Type
Giants   147

P. DUBOIS, M. JASCHEK, and C. JASCHEK / Spectral Classification in the Small
Magellanic Cloud   149

N. R. WALBORN / Nitrogen and Carbon Anomalies in OB Spectra   153

Y. ANDRILLAT and L. HOUZIAUX / Infrared Lines in Peculiar Emission-Line
Stars   155

C. VAN 'T VEER and C. BURKHART / Conflicting Evidence on the Composition
of Am Stars   157

K. NANDY, G. I. THOMPSON, and C. M. HUMPHRIES / Ultraviolet Spectrophotometry of Early-Type Stars   161

M.-N. PERRIN / Mass Estimation of Twelve K Dwarfs   167

L. E. PASINETTI / Can we Check by High-Dispersion Analyses Stars Belonging to Eggen's Moving Groups?   173

A. CUCCHIARO, M. JASCHEK, C. JASCHEK, and D. MACAU-HERCOT / Spectral Classification of B and A Stars from Data of S2/S68 Experiment   177

PART V / ABUNDANCES IN STELLAR POPULATIONS

H. SPINRAD *(Invited Lecture)*   183

W. P. BIDELMAN and W. G. SMETHELLS / A Population Discriminant in M-Dwarf Spectra   205

M. MAYOR / Chemical Evolution of the Galactic Disk and the Radial Metallicity Gradient   207

R. FOY / Abundances in the Galactic Clusters of Hyades and M 67   209

R. GRIFFIN / Curve-of-Growth Analysis of a Red Giant in M 67   213

S. GRENIER, L. DA SILVA, and A. HECK / Some Comments on the Age-Abundance Relation   215

W. W. MORGAN / Remarks on Some Aspects of Spectral Classification   219

APPENDIX / A CATALOGUE OF [Fe/H] DETERMINATIONS

M. MOREL, C. BENTOLILA, G. CAYREL, and B. HAUCK / A Catalogue of [Fe/H] Determinations   223

INDEX OF NAMES   261

PREFACE

The general discussions of the roles of photometric and spectroscopic classification at Cordoba in 1971 (IAU Symposium No. 50), and of the calibration of classification indices at Geneva in 1972 (IAU Symposium No. 54), revealed clearly the steadily increasing importance of abundance parameters. The multiplicity of these, however, raised so many new problems that it was logical that the 1975 meeting at Lausanne should be concerned with ways in which differences in abundance affect both spectral types and photometric indices. Commissions 29 and 36 joined with Commission 45 in sponsoring this Symposium. Since the date of the meeting came shortly after the formal retirement of Professor William W. Morgan from the University of Chicago, it was quickly agreed that this meeting should be dedicated to him in recognition of his unique contributions to spectral classification. In the opening paper of the Symposium Dr. Bengt Strömgren has summarized these. To his remarks we should add only that it was about 1940 that Morgan first distinguished the group of G- and K-type stars with weak CN bands and metallic lines — stars which have since been recognized as having the abundance of all metals relative to hydrogen much lower than in stars of the solar population. Spectra of two of these, HD 81192 (Boss 2527) and δ Lep, were later shown as examples of the group in the Yerkes Atlas of 1943.

The topic of this Symposium can be thought of as involving two somewhat different problems. The first is the effect of abundance differences on the various indices used to measure the fundamental physical variables, temperatures and luminosity. The second problem is that of recognizing the number of groups of elements which may vary independently from star to star, and then of measuring these differences. The papers contained in this volume suggest that although progress has been made on both problems, neither is close to complete solution.

The hosts for this Symposium were the Institute of Astronomy of the University of Lausanne and the Geneva Observatory, with the sessions held in the modern College Propédeutique of the University of Lausanne in the suburb of Dorigny.

The success of the meeting was due both to the careful scientific planning of Dr Carlos Jaschek, Chairman of Commission 45, and to the exhausting work of the Local Organizing Committee. Financial assistance was provided by the Executive Committee of the I.A.U., the *Fonds national suisse de la recherche scientifique*, the Swiss Academy of Sciences and the Universities of Lausanne and Geneva. Mrs B. Michaud and Mrs B. Wilhelm are gratefully thanked for their assistance at all stages of preparation of the present volume.

THE EDITORS

SCIENTIFIC ORGANIZING COMMITTEE

B. Baschek, W. P. Bidelman, M. Golay, M. Hack, B. Hauck,
C. Jaschek (Chairman), P. C. Keenan, B. Strömgren

LOCAL ORGANIZING COMMITTEE

M. Golay, B. Hauck (Chairman), P. Javet, B. Junod, A. Maeder, F. Rufener

LIST OF PARTICIPANTS

J. Andersen, Copenhagen

H. Andrillat, Saint Michel

Y. Andrillat, Saint Michel

C. Arpigny, Liège

B. Baschek, Heidelberg

A. Behr, Geneva

R. Bell, Maryland

W. P. Bidelman, East Cleveland

P. Bouvier, Geneva

I. Bues, Berlin

C. Burkhart, Lyon

G. Burki, Geneva

R. Canavaggia, Paris

F. Castelli, Trieste

G. Cayrel de Strobel, Meudon

B. Cester, Trieste

D. Chalonge, Paris

L. E. Cram, Paris

N. Cramer, Geneva

D. Crawford, Tucson

A. Cucchiaro, Liège

L. da Silva, Rio de Janeiro

J. Delhaye, Meudon

R. de la Reza, Geneva

P. Dubois, Strasbourg

D. Egret, Strasbourg

R. Foy, Meudon

M. Fracassini, Milan

I. Furenlid, Tucson

R. F. Garrison, Ontario

M. Gerbaldi, Paris

M. Golay, Geneva

S. Grenier, Meudon

M. Grenon, Geneva

R. E. M. Griffin, Cambridge, U.K.

B. Gustafsson, Uppsala

K. Gyldenkerne, Tølløse

M. Hack, Trieste

B. Hauck, Lausanne

A. Heck, Liège

N. Houk, Ann Arbor

L. Houziaux, Mons

C. Jaschek, Strasbourg

M. Jaschek, Strasbourg

P. Javet, Lausanne

B. Junod, Geneva

R. Kandel, Meudon

P. C. Keenan, Delaware

F. Llorente de Andres, Geneva

K. Lodén, Stockholm

P. B. Lucke, Geneva

A. Maeder, Geneva

P. Magnenat, Lausanne

N. Mandwewala, Geneva

P. Martin-Mullor, Paris

M. McCarthy, Vatican City

E. E. Mendoza V., Mexico

J.-Cl. Mermilliod, Lausanne

M. Morel, Lausanne

W. W. Morgan, Williams Bay

N. Morguleff, Paris

E. E. Müller, Geneva

K. Nandy, Edinburgh

C. Nicollier, Lausanne

B. Nicolet, Lausanne

P. E. Nissen, Aarhus

B. Nordström, Copenhagen

E. Oblak, Besançon

T. Oja, Kvistaberg

E. H. Olsen, Copenhagen

W. Osborn, Mérida

L. E. Pasinetti, Milan

C. Payne-Gaposchkin, Cambridge, U.S.A.
H. Pedersen, Aarhus
M.-N. Perrin, Meudon
D. Proust, Meudon
A. Przybylski, Mount Stromlo
F. Querci, Meudon
M. Querci, Meudon
A. Reisz, Copenhagen
M. Rudkjobing, Aarhus
F. Rufener, Geneva
A. J. Sauval, Brussels
E. Schatzman, Meudon
W. Seggewiss, Bonn

W. C. Seitter, Bonn
H. Spinrad, Berkeley
U. Steinlin, Basle
C. B. Stephenson, East Cleveland
A. Strobel, Torun
B. Strömgren, Copenhagen
A. Unsöld, Kiel
C. van 't Veer, Paris
N. R. Walborn, La Serena
R. Wehrse, Heidelberg
P. M. Williams, Edinburgh
B. Wolf, Santiago

WELCOME TO W. W. MORGAN

B. STRÖMGREN

Copenhagen Observatory, Denmark

It is a great privilege to have the opportunity to welcome William Morgan to the IAU Symposium that has been opened today.

All those present are familiar in a general way with William Morgan's outstanding contributions to modern Astronomy and Astrophysics. During the last several weeks I have reread a number of his papers, recalling the situation in Astronomy at the time when they were written, and also reconsidering them in the light of what we know today. It has been a most rewarding experience, and I should like to share my impressions with you.

William Morgan's first papers appeared during the years 1927–1931, and from 1932 on the *Astrophysical Journal* contains a large number of articles and notes by Morgan, mostly containing spectroscopic results pertaining to peculiar stars and spectrum variables. In 1933 the *Astrophysical Journal* contained two important papers by Morgan on problems of spectral classification. I wish to quote the titles, namely, 'Some Effects of Changes in Stellar Temperature and Absolute Magnitude', and 'Some Evidence for the Existence of a Peculiar Branch of the Spectral Sequence in the Interval B8–F0'. Already at this time William Morgan was contributing significantly to the field with which this IAU Symposium is concerned.

In 1935 followed his monumental detailed investigation of A stars, 'A Descriptive Study of the Spectra of the A-Type Stars'. This publication contained estimated intensities for ten important absorption lines in the spectra of 125 stars brighter than magnitude 5.5 as well as detailed spectroscopic information, including estimated intensities, on several hundred absorption lines in the spectra of thirteen selected A stars. The material published in this work greatly furthered the understanding of the problem of A stars, and the deep familiarity with spectroscopic material that Morgan had gained through this research must have been an invaluable asset in his later investigations. One of the selected thirteen stars was 15 UMa, described as being composite in the Henry Draper Catalogue. Morgan demonstrated quite clearly that the peculiarities could not be explained through this assumption, and his analysis foreshadows his later highly important work on metallic line stars.

Before referring to Morgan's following papers on spectral classification I wish to mention a note in the *Astrophysical Journal* entitled 'A useful Fine-Grain Developer for Spectrographic Photography'. In all his spectroscopic work William Morgan took great pains to secure the highest possible quality of the observational material, attending to every detail of the process of obtaining the stellar spectra.

There followed in 1937 the paper, 'On the Spectral Classification of the Stars of Types A to K', and in 1938 'On the Spectral Types and Luminosities of the M Dwarfs', and 'On .

B. Hauck and P. C. Keenan (eds.), Abundance Effects in Classification, XV– XVII. *All Rights Reserved. Copyright* ⨃ *1976 by the IAU.*

the Determination of Color Indices of Stars from a Classification of Their Spectra'. With J. Titus, Morgan published in 1940 a paper on the spectral types of the brighter members of the Hyades cluster, and in the same year with P. C. Keenan an article entitled 'The Classification of the Red Carbon Stars'.

By this time the essential features of MK classification must have already been almost completely clear. The system of MK classification was definitely established in 1943 with the publication of the classical contribution by W. W. Morgan, P. C. Keenan and E. Kellman, 'An Atlas of Stellar Spectra with an Outline of Spectral Classification'. Few publications have had a comparable influence, and in few cases has similar definitiveness been achieved in the first great attempt.

It is of interest to consider the background in the field of stellar classification. The one-dimensional Harvard classification had been of great importance in stellar astrophysics and in galactic research, and the Mount Wilson work on determination of absolute magnitudes and stellar distances from spectra, initiated by Adam and Kohlschütter, had had a great impact, although at the time considered there had also been some disappointments.

A number of features of the MK system of two-dimensional classification contributed to its success. Suffice it to emphasize the importance of the introduction of classification based on comparison spectra of standard stars, selected with great care and on the basis of extensive experience; and the fortunate choice of luminosity classes, the M_v-calibration of which was to follow and to be improved through continued research.

Friends of William Morgan, who over the years have been privileged to discuss problems of stellar classification with him have been greatly impressed by the enormous amount of information on stellar spectra that he carried in his head. I have myself also been particularly struck by the fine balance in his whole approach, with proper emphasis on the importance − in stellar astrophysics as well as in galactic research − of classification of the great bulk of normal stars, as well as on the importance of segregation, and further intensive study, of deviating, or peculiar stars.

The subject of deviating stars was pursued in the important paper by N. G. Roman, W. W. Morgan and O. J. Eggen, 'The Classification of Metallic Line Stars', published in 1948. Let me also mention here the article by P. C. Keenan, W. W. Morgan and G. Münch, 'Spectra of High-Velocity Giants'.

I now wish to turn to William Morgan's investigations in galactic research, based on the use of MK classification, and on further developments with a particular view to high-luminosity stars.

In 1949 Morgan, with J. J. Nassau, published an investigation based on the use of Schmidt astrograph plates entitled 'A Survey of Stars of High Luminosity in the Galaxy'. This was followed in 1951 by the paper by J. J. Nassau and W. W. Morgan, 'A Finding List of O and B Stars of High Luminosity', and in 1952 by 'A Finding List of F Stars of High Luminosity', by the same authors; and William Morgan described his method of classification of natural groups, so useful in selecting candidates for further detailed spectroscopic studies, in *Publ. Obs. Univ. Michigan*, Vol. 10.

The application of the new methods of galactic research led to a striking result in

1952. W. W. Morgan, S. Sharpless and D. Osterbrock published the paper 'Some Features of Galactic Structure in the Neighborhood of the Sun'. Through judicious selection of galactic objects, namely very young clusters and associations, and careful determination of spectroscopic distances, the existence of spiral arm structure in our Galaxy was established. Later work in optical astronomy, notably by Wilhelm Becker and his collaborators, confirmed and enlarged these findings, and fully confirmed the soundness of the approach by Morgan and his collaborators.

Radio-astronomical investigations of the spiral structure of our Galaxy have contributed greatly, but the problem is complex, and more and more it has become clear that these investigations should be combined with those of optical astronomy. The 1952 investigation on spiral structure of Morgan and his collaborators is certainly pioneer work of the highest importance.

In collaboration with A. D. Code and A. E. Whitford, W. W. Morgan continued work in galactic research. Photoelectric photometry was combined with MK classification to yield most valuable results on the distribution of early-type stars, with a particular view to associations.

Fundamental work in stellar astrophysics was carried out by William Morgan in collaboration with H. L. Johnson. The powerful methods of MK classification and *UBV* photometry were combined in an investigation on standard stars defining spectral types, published in 1953. The paper in question became a standard reference work of great importance in subsequent research.

In numerous papers, mostly based on joint investigations with other astronomers, Morgan has continued galactic research.

In more recent years William Morgan has turned his attention to problems of external galaxies. His vast experience in the field of stellar classification was brought to bear on questions of classification of the highly composite spectra of, first globular clusters, and then external galaxies. The results have been of great importance in the analysis of problems of external galaxies. In particular, parallel investigations of the morphology of external galaxies have led to a considerably deepened understanding of the problems of galaxy classification.

Of particular importance to the astrophysical discussion of today is William Morgan's research on the classification of compact galaxies. Results of studies of morphology and spectra of compact galaxies have led to important new insights, and again William Morgan's power of combination of great amounts of observational information has been the basis of success.

Quite recently William Morgan has tackled a problem characterized by still larger dimensions, namely, the classification of clusters of galaxies, and already significant results have been published.

I have dwelt particularly on William Morgan's contributions in the field of stellar classification; for it is above all these contributions that come to mind in connection with the subject of this IAU Symposium.

I am sure that you agree that it was a wonderfully justifiable decision to honour him by calling it: The William Morgan Symposium of the IAU.

BIBLIOGRAPHY OF W. W. MORGAN

1927 (with O. Struve)	Orbit of the Spectroscopic Binary 95 o Leonis, *Astrophys. J.* **66**, 135.
(Abstract)	Spectroscopic and Photometric Observations of CP Cygni, *Pop. Astron.* **35**, 490.
	Observations of the Short-Period Variable AK Herculis, *Pop. Astron.* **35**, 547.
1928 (Abstract)	Two Recent Solar Eruptions, *Pop. Astron.* **36**, 604.
1930	Leonids Observed at the Yerkes Observatory, *Pop. Astron.* **38**, 624.
	Light-Curve and Orbit of the Eclipsing Binary ZZ Aurigae, *Pop. Astron.* **38**, 466.
1931	Studies in Peculiar Stellar Spectra. I. The Manganese Lines in α Andromedae, *Astrophys. J.* **73**, 104.
	Studies in Peculiar Stellar Spectra. II. The Spectrum of B. D. $-18°3789$, *Astrophys. J.* **74**, 24.
1932 (with O. Struve)	On the Intensities of Stellar Absorption Lines, *Proc. Nat. Acad. Sc.* Washington **18**, 590.
	Studies in Peculiar Stellar Spectra. III. On the Occurrence of Europieum in A-Type Stars, *Astrophys. J.* **75**, 46.
	On the Variable Lines of Helium and Magnesium in the Spectrum of 13μ Sagittarii, *Astrophys. J.* **75**, 407.
	Note on the Spectrum of 37θ Aurigae, *Astrophys. J.* **75**, 423.
	A Study of the Composite Spectrum of the A-Type Star 14 Comae, *Astrophys. J.* **76**, 144.
	Four Early-Type Stars Having Variable Absorption Lines, *Astrophys. J.* **76**, 275.
(with G. Farnsworth)	The Spectrum of the A2 Dwarf ϵ Serpentis, *Astrophys. J.* **76**, 299.

B. Hauck and P. C. Keenan (eds.), Abundance Effects in Classification, XIX—XXVIII. All Rights Reserved.
Copyright © 1976 by the IAU.

Four A- and F-Type Stars Whose Spectra Contain Variable Absorption Lines, *Astrophys. J.* **76**, 315.

1933 (with O. Struve and C. T. Elvey)   Note on the Occultation of BD + 8°2456 by Jupiter, *J. Brit. Astron. Assoc.* **43**, 325.

A Study of the Spectrum Variable 73 Draconis, *Astrophys. J.* **77**, 77.

On a Line at λ 4470 in the Spectrum of 21 Aquilae, *Astrophys. J.* **77**, 226.

Some Effects of Changes in Stellar Temperatures and Absolute Magnitudes, *Astrophys. J.* **77**, 291.

Some Evidence for the Existence of a Peculiar Branch of the Spectral Sequence in the Interval B8-F0, *Astrophys. J.* **77**, 330.

On the Relative Intensities of Certain Lines of Ti II in the Spectra of Three F-Type Stars, *Astrophys. J.* **78**, 158.

1934   A II in the Spectrum of υ Sagittarii, *Astrophys. J.* **79**, 513.

(with B. A. Wooten)   Relative Stellar Energy Distribution in the Infrared, *Astrophys. J.* **80**, 229.

1935   A Descriptive Study of the Spectra of The A-Type Stars, *Publ. Yerkes Obs.* **7**, 133.

A Peculiar Spectroscopic Phenomenon in The Algol System, *Astrophys. J.* **81**, 348.

(with O. Struve and C. T. Elvey)   Variations in the Spectrum of 29 Canis Majoris, *Astrophys. J.* **82**, 95.

1936   A Possible Interpretation of the Absorption Spectra of Nova Herculis, *Astrophys. J.* **83**, 252.

A Useful Fine-Grain Developer for Spectrographic Photography, *Astrophys. J.* **83**, 254.

1937   On the Spectral Classification of the Stars of Types A to K, *Astrophys. J.* **85**, 380.

Note on a Possible Effect of the Recent Increase in Brightness of γ Cassiopeiae, *Astrophys. J.* **86**, 100.

1938   On the Determination of Color Indices of Stars From a Classification of Their Spectra, *Astrophys. J.* **87**, 460.

On the Spectral Types and Luminosities of

the M. Dwarfs, *Astrophys. J.* **87**, 589.

(with C. T. Elvey)   A New Fourth-Magnitude Eclipsing Binary, *Astrophys. J.* **88**, 110.

Storrs B. Barrett, *Pop. Astron.* **46**, 127.

1939   Molecular Bands as Indicators of Stellar Temperatures and Luminosities, *Astrophys. J.* **89**, 310.

(with F. Sherman)   On the Color of P Cygni, *Astrophys. J.* **89**, 509.

(with F. Sherman)   The Nebulosity near S Monocerotis, *Astrophys. J.* **89**, 553.

(with P. C. Keenan)   On the Spectral Type of the Sun, *Publ. Astron. Soc. Pacific* **51**, 355.

Note on Interstellar Reddening in the Region of γ Cygni, *Astrophys. J.* **90**, 632.

(with W. J. Luyten and O. Struve)   Re-observation of the Orbits of Ten Spectroscopic Binaries with a Discussion of Apsidal Motions, *Publ. Yerkes Obs.* **7**, IV, 251.

1940 (with J. Titus)   On the Classification of the A Stars. I. The Spectral Types of the Brighter Members of the Hyades Cluster, *Astrophys. J.* **92**, 256.

1941 (with P. C. Keenan)   The Classification of the Red Carbon Stars, *Astrophys. J.* **94**, 501.

Yerkes Observatory (1897–1941) and Some Notes on the Astronomical Struves, *The Sky* **5**, No. 8.

Some Uses of Spectral Classification, *The Telescope* **8**, 77 and 92.

1943 (with P. C. Keenan and E. Kellman)   *An Atlas of Stellar Spectra, With An Outline of Spectral Classification*, Univ. of Chicago Press.

(Abstract)   On the Parallaxes of Some of the Brighter Helium Stars, *Publ. Amer. Astron. Soc.* **10**, 310.

1944 (with W. A. Hiltner)   Complex Lines in the Spectrum of θ Aurigae, *Astrophys. J.* **99**, 318.

(Abstract)   The Helium Stars Having the Most Rapid Axial Rotation, *Astron. J.* **51**, 21.

(Abstract)   Note on Interstellar Reddening in the Region of the Orion Nebula, *Astron. J.* **51**, 21.

1946 (with W. P. Bidelman)   New Hα Emission Stars, *Astrophys. J.* **103**, 378.

(with I. Hansen) Be Stars Showing Bright Lines of Fe II *Astrophys. J.* **103**, 379.

(with S. Sharpless) An Interesting Emission-Line Star Near the Orion Nebula, *Astrophys. J.* **103**, 249.

(with W. P. Bidelman) On the Interstellar Reddening in the Region of the North Polar Sequence and the Normal Color Indices of A-Type Stars, *Astrophys. J.* **104**, 245.

1947 (with G. P. Kuiper) Studies of the Stellar System, *Science* **106**, 201.

(with A. J. Deutsch) The Spectrum of T Coronae Borealis at its 1946 Outburst, *Astrophys. J.* **106**, 362.

1948 (with P. C. Keenan and G. Munch) (Abstract) Spectra of High-Velocity Giants, *Astron. J.* **53**, 194.

(with N. G. Roman and O. J. Eggen) The Classification of the 'Metallic-Line' Stars, *Astrophys. J.* **107**, 107.

1949 (with J. J. Nassau) (Abstract) A Survey for Stars of High Luminosity in the Galaxy, *Astron. J.* **54**, 192.

1950 (with N. G. Roman) The Moving Cluster in Perseus, *Astrophys. J.* **111**, 426.

(with N. G. Roman) Revised Standards for Supergiants on the System of the Yerkes Spectral Atlas, *Astrophys. J.* **112**, 362.

(with J. J. Nassau) A Finding List of O and B Stars of High Luminosity, *Astrophys. J.* **113**, 141.

1951 Application of the Principle of Natural Groups to the Classification of Stellar Spectra, *Publ. Obs. Univ. Michigan* **10**, 33.

(with J. J. Nassau) Distribution of Early Type Stars of High Luminosity Near the Galactic Equator, *Publ. Obs. Univ. Michigan* **10**, 43.

(with H. L. Johnson) On the Color-Magnitude Diagram of the Pleiades, *Astrophys. J.* **114**, 522.

(with W. S. Fitch) A Useful Luminosity Discriminant in the Late K and Early M Giants, *Astrophys. J.* **114**, 548.

(with P. C. Keenan) *Classification of Stellar Spectra*, Ch. 1 in *Astrophysics* (ed. by J. A. Hynek), McGraw-Hill.

1952 (with J. J. Nassau) A Finding List of F Stars of High Luminosity, *Astrophys. J.* **115**, 475.

(with S. Sharpless and
D. Osterbrock)
(Abstract)

Some Features of Galactic Structure in the
Neighborhood of the Sun, *Astron. J.* **57**, 3.

(with A. D. Code and
A. E. Whitford)
(Abstract)

The Reddest B Stars, *Astron. J.* **57**, 8.

(with J. J. Nassau and V. Blanco)
(Abstract)

Surface Distribution of Late M Stars in the
Milky Way, *Astron. J.* **57**, 21.

1953 (with A. Blaauw)

Expanding Motions in the Lacerta Aggre-
gate, *Astrophys. J.* **117**, 256.

(with H. L. Johnson)

Fundamental Stellar Photometry for Stan-
dards of Spectral Type on the Revised
System of the Yerkes Spectral Atlas,
Astrophys. J. **117**, 313.

(with G. Haro)

Rapid Variables in the Orion Nebula,
Astrophys. J. **118**, 16.

(with D. L. Harris and
H. L. Johnson)

Some Characteristics of Color Systems,
Astrophys. J. **118**, 92.

(with L. Munch)

A Probable Clustering of Blue Giants in
Cygnus, *Astrophys. J.* **118**, 161.

(with A. E. Whitford and
A. D. Code)

Studies in Galactic Structure. I. A Preliminary Determination of the Space Distribution
of the Blue Giants, *Astrophys. J.* **118**, 318.

(with Gu. Gonzalez and
Gr. Gonzalez)

Blue Giants in the Neighborhood of NGC
6231, *Astrophys. J.* **118**, 323.

(with Gu. Gonzalez and
Gr. Gonzalez)

Blue Giants in the Direction of the Galactic
Center, *Astrophys. J.* **118**, 345.

(with A. Blaauw)

Note on the Motion and Possible Origin of
the O-Type Star HD 34078 = AE Aurigae
and the Emission Nebula IC 405, *Bull.
Astron. Inst. Neth.* **448**, 76.

1954 (with H. L. Johnson and
N. G. Roman)

A Very Red Star of Early Type in Cygnus,
Publ. Astron. Soc. Pacific **66**, 85.

Some Astrometric Problems of Galactic
Structure, *Astron. J.* **59**, 86.

(with B. Stromgren and
H. M. Johnson)
(Abstract)

New Features of Some Emission Regions in
the Milky Way, *Astron. J.* **59**, 188.

(with H. L. Johnson)

A Heavily Obscured O-Association in
Cygnus, *Astrophys. J.* **119**, 344.

(with D. L. Harris and
N. G. Roman)

Photometric and Spectroscopic Observations
of Stars in IC 348, *Astrophys. J.* **119**, 622.

(with A. Blaauw)

The Space Motions of AE Aurigae and μ

Columbae with Respect to the Orion Nebula, *Astrophys. J.* **119**, 625.

(with J. J. Nassau and V. M. Blanco)
Reddened Early M- and S-Type Stars Near the Galactic Equator, *Astrophys. J.* **120**, 478.

(with A. B. Meinel and H. M. Johnson)
Spectral Classification with Exceedingly Low Dispersion, *Astrophys. J.* **120**, 506.

1955
La Classification Spectrale de deux Populations Stellaires, In *Principes Fondamentaux de Classification Stellaire, Coll. Int. C.N.R.S.* (June–July, 1953), **55**, 7.

(with A. Blaauw and F. C. Bertiau)
19 Scorpii, a Luminous A Star in the Scorpio-Centaurus Association, *Astrophys. J.* **121**, 557.

(with B. Stromgren and H. M. Johnson)
A Description of Certain Galactic Nebulosities, *Astrophys. J.* **121**, 611.

(with A. D. Code and A. E. Whitford)
Studies in Galactic Structure. II. Luminosity Classification for 1270 Blue Giant Stars, *Astrophys. J. Suppl.* **2**, 41.

(with H. L. Johnson)
Some Evidence for a Regional Variation in the Law of Interstellar Reddening, *Astrophys. J.* **122**, 142.

(with H. L. Johnson)
Photometric and Spectroscopic Observations of the Double Cluster in Perseus, *Astrophys. J.* **122**, 429.

Spiral Structure of the Galaxy, *Scientific American* **192**, 42.

1956
The Integrated Spectral Types of Globular Clusters, *Publ. Astron. Soc. Pacific* **68**, 509.

(with D. L. Harris)
The Galactic Cluster M29 (NGC 6913), *Vistas in Astronomy* **2**, 1124.

1957 (with N. L. Gould and G. H. Herbig)
BD +75°325: A Subluminous O-Type Star, *Publ. Astron. Soc. Pacific* **69**, 242.

(with N. U. Mayall)
A Spectral Classification of Galaxies, *Publ. Astron. Soc. Pacific* **69**, 291.

(with H. A. Abt and B. Stromgren)
A Description of Certain Galactic Nebulosities. II, *Astrophys. J.* **126**, 322.

1958
A Preliminary Classification of the Forms of Galaxies According to their Stellar Population, *Publ. Astron. Soc. Pacific* **70**, 364.

The Observation of Blue Giant Stars at Great Distances from the Sun, In *The Large-Scale*

	Structure of the Galactic System, IAU Symp. **5**, 57.
	Systems of Spectroscopic Luminosity Criteria, *Astron. J.* **63**, 180.
	Some Characteristics of the Strong- and Weak-Line Stars, In Stellar Populations, Conference, Pontifical Acad. Sci., Vatican City, May 1958, *Specola Vat. Richerce Astron.* **5**, 323.
	Some Features of Stellar Populations as Determined From Integrated Spectra. *Ibid.*, 325.
	Some Spectroscopic Phenomena Associated with the Stellar Population of Galaxies. In *La Structure de l'Evolution de l'Universe*, 11th Solvay Physics Conf. Brussels, p. 297.
(Abstract)	On the Stellar Population of the Nuclear Region of the Galaxy. Read at National Acad. Sciences, Nov., 1958, *Science* **128**, 1147.
1959	The Spectra of Galaxies, *Publ. Astron. Soc. Pacific* **71**, 92.
	A Preliminary Classification of the Forms of Galaxies According to Their Stellar Population, *Publ. Astron. Soc. Pacific* **71**, 352.
	The Integrated Spectra of Globular Clusters, *Astron. J.* **64**, 432.
(Book review)	F. Zwicky: Morphological Astronomy, *Publ. Astron. Soc. Pacific* **71**, 352.
1960	Some Vistas of Astronomical Discovery, *Science* **132**, 73 (Reprinted in *Publ. Astron Soc. Pacific* **72**, 153).
	Some Characteristics of Galaxies Which Bear on Their Use as a Fundamental Astrometric Frame of Reference, *Astron. J.* **65**, 222.
1961	The Classification of Clusters of Galaxies, *Proc. National Acad. of Sciences*, Washington, **47**, 905.
1962	Some Characteristics of Galaxies, (The Henry Norris Russell lecture), *Astrophys. J.* **135**, 1.
(with T. A. Mathews)	On the Optical Forms of the Brightest Extra-

	(Abstract)	galactic Radio Sources, *Science* **138**, 992.
	(with N. U. Mayall)	Spectral Types of Galaxies, In 'Symposium on Problems of Extra-Galactic Research', *IAU Symp.* **15** (Santa Barbara), 22.
1963	(with W. A. Hiltner)	Spectroscopic Discriminant for O-Type Star Clusters, *Astron. J.* **68**, 281.
	(Abstract)	
	(with J. S. Neff)	Color Separation of Giant and Dwarf Stars, *Astron. J.* **68**, 288.
	(Abstract)	
1964	(with T. A. Mathews and M. Schmidt)	A Discussion of Galaxies Identified With Radio Sources, *Astrophys. J.* **140**, 35.
	(with J. S. Neff)	Stellar Population of the Nuclear Region of the Andromeda Nebula, *Astron. J.* **69**, 145.
	(Abstract)	
1965	(with W. A. Hiltner)	Studies in Spectral Classification. I. The HR Diagram of the Hyades, *Astrophys. J.* **141**, 177.
	(with W. A. Hiltner and J. S. Neff)	Studies in Spectral Classification. II. The HR Diagram of NGC 6530, *Astrophys. J.* **141**, 183.
	(with W. A. Hiltner, J. S. Neff, R. F. Garrison and D. E. Osterbrock)	Studies in Spectral Classification. III. The HR Diagram of NGC 2244 and NGC 2264, *Astrophys. J.* **142**, 974.
	(with J. Lesh)	The Supergiant Galaxies, *Astrophys. J.* **142**, 1364.
1966		Introductory Paper. In 'Spectral Classification and Multicolor Photometry', *IAU Symp.* **24**, Saltsjöbaden, Aug. 1964, 3.
	(with K. Lodén)	Some Characteristics of the Orion Association, *Vistas in Astronomy* **8**, 83.
1967		Frank Elmore Ross. *Biographical Memoirs Natl. Acad. Sci.*, Washington, **39**.
		A Note on the MK Classification System, In *Modern Astrophysics. A memorial to Otto Struve* (ed. by M. Hack), Paris, Gauthier-Villars, p. 83.
1968	(with H. J. Smith and D. Weedman)	Spectral Variations in the Galaxy NGC 4151, *Astrophys. J. Letters* **152**, L113.
		A Comparison of the Optical Forms of Certain Seyfert Galaxies with the N-Type Radio Galaxies, *Astrophys. J.* **153**, 27.
	(with C. Jaschek, M. Jaschek and A. Slettebak)	On the Spectrum of Bidelman's Helium Variable Star HD 125823, *Astrophys. J. Letters* **153**, L87.

(with H. A. Abt, A. B. Meinel and J. W. Tapscott)	*An Atlas of Low-Dispersion Grating Stellar Spectra,* Kitt Peak Nat. Obs., Tucson.
	Large Scale Distribution of H II Regions in Galaxies, In *Interstellar Ionized Hydrogen* (ed. by Y. Terzian), W. A. Benjamin, N. Y., 565.
1969 (with D. E. Osterbrock)	On the Classification of the Forms and the Stellar Content of Galaxies, *Astron. J.* **74**, 515.
(with H. A. Abt)	The HR Diagram of NGC 2516. *Astron. J.* **74**, 813.
(with W. A. Hiltner)	UBV Photometry and Spectral Types in NGC 6611, *Astron. J.* **74**, 1152.
1970	Struve's Approach to Spectral Classification, In *Spectroscopic Astrophysics* (ed. by G. H. Herbig), Univ. of California Press, Berkeley, p.25.
	Spiral Structure in External Galaxies, In 'The Spiral Structure of Our Galaxy', *IAU Symp.* **38** (Basel), 9.
(with L. P. Bautz)	On the Classification of the Forms of Clusters of Galaxies, *Astrophys. J.* **162**, L149.
1971 (with D. S. Heeschen and N. R. Walborn)	The Distribution of Light in the Central Regions of Some Giant Elliptical Galaxies, *Astrophys. J.* **165**, L65.
(with W. A. Hiltner and R. F. Garrison)	The HR Diagram of the Alpha Persei Cluster, *Astron. J.* **76**, 242.
(with N. R. Walborn and J. W. Tapscott)	An Optical Form Morphology of Seyfert Galaxies, In Study Week on Nuclei of Galaxies (April 13–18, 1970). (ed. by D. J. O'Connell), *Pontif. Acad. Sci. Scripta Varia* **35**, 27.
	A Unitary Classification for N Galaxies, *Astron. J.* **76**, 1000.
1972	Classification of Compact Objects: QSS, QSO's, N-Type and Compact Galaxies, Seyfert and Galactic Nuclei, 'In External Galaxies and Quasi-Stellar Objects'. (ed. by D. S. Evans). Uppsala, Aug. 1970, *IAU Symp.* **44**, 97.
(with H. A. Abt)	The Spectral Classification of the F Stars of

	(with H. A. Abt)	Intermediate Luminosity, *Astron. J.* **77**, 35. The HR Diagram of the Open Cluster IC 2602, *Astrophys. J.* **174**, L131.
1973	(with R. A. White and J. W. Tapscott)	A New Shell Phase in Pleione, *Astron. J.* **78**, 302.
	(with H. A. Abt)	On the Metallicity of the Main-Sequence Stars in M67, *Astron. J.* **78**, 386.
	(with P. C. Keenan)	Spectral Classification, *Ann. Rev. Astron. Astrophys.* **11**, 29.
1975	(with S. Kayser and R. A. White)	cD Galaxies in Poor Clusters. (in press).
	(with H. A. Abt)	MK Morphology of a Group of Am Stars, *Astrophys. J.* (in press). Remarks on Some Aspects of Spectral Classification, *IAU Symp.* **72**.

PART I

INFLUENCE OF ABUNDANCES UPON
STELLAR ATMOSPHERE CALCULATIONS

INFLUENCE OF ABUNDANCES UPON STELLAR ATMOSPHERE CALCULATIONS

B. BASCHEK

Lehrstuhl für Theoretische Astrophysik, Universität Heidelberg, Germany

Abstract. The basic equations for constructing a stellar atmosphere (hydrostatic equilibrium, flux constancy, radiative transfer, convective instability) are briefly summarized. While the parameters T_{eff} (effective temperature) and g (surface gravity) are directly contained in these equations, the element abundances ϵ_i enter only indirectly through the thermodynamic properties (such as electron pressure, entropy, ...) and the absorption and scattering coefficients of stellar matter.

The equation of state, convection, the effects of the absorption coefficients (particularly of line absorption) on the temperature stratification, and the role of velocity fields (microturbulence) are discussed in some detail, emphasizing their dependence on the abundances.

From a given model atmosphere, a 'theoretical spectrum' (colours, bolometric correction, line strengths etc.) can be calculated. The (relative) fluxes emerging at the surface are essentially determined by the temperature gradient and the absorption coefficients at the frequencies under consideration. The basic goal of quantitative classification, however, is the inverse problem, namely to deduce the stellar parameters from selected observed spectral criteria. Aspects relevant to this problem such as the question of uniqueness and the occurrence of possible systematic errors (even when using differential analysis techniques) are briefly sketched and illustrated by some examples.

1. Introduction

Spectral classification is based on the observation of selected spectral criteria (such as intensity ratios of lines, colours etc.) which are compared with those of a set of standard stars defining the classification system. The ultimate goal is to derive the basic stellar parameters, including the abundances of the elements. This requires a quantitative calibration of the criteria in terms of these parameters so that observed spectra can be compared with theoretical spectra. The most detailed and reliable procedures to achieve quantitative spectral classification make use of model atmosphere techniques.

As this Symposium is devoted to the discussion of abundance effects in classification, it seems useful to give a brief summary of the theoretical background for calculating stellar atmospheres with particular emphasis on the influence of the *element abundances* upon the atmospheric structure. In order to illustrate the basic effects of the abundances no detailed discussion is required. It suffices to use fairly simple – long known – approximations. Of course, any actual reliable calculation of theoretical spectra should involve very elaborate models. Details of the physical and numerical problems of stellar atmospheres can be found, for example, in the books by Mihalas (1970) or by Unsöld (1955). Effects of abundances have recently been considered by Pecker (1973) in a discussion of the use of model atmospheres for temperature-gravity calibration.

In the following, the basic equations governing the structure of the atmosphere will be summarized, followed by a discussion of the effects of the abundances on the structure (through the thermodynamic properties and the absorption of stellar matter). Finally, the relationship between theoretical and observed spectra and some other aspects relevant to spectral classification will be briefly discussed.

2. Basic Parameters and Equations

We consider a 'standard model atmosphere' which is assumed to be homogeneous, plane-parallel, in hydrostatic equilibrium with constant (radiative plus convective) flux. Furthermore, local thermodynamic equilibrium (LTE) is assumed to hold (continuous scattering may be allowed for, but is not included here). The atmospheric structure and hence the emergent radiation is then determined by the parameters (some 94, in principle):

(i) effective temperature T_{eff}
(ii) surface gravity g, and
(iii) relative element abundances ϵ_i (usually expressed as atomic numbers, normalized to $\log \epsilon_H = 12.00$).

The structure of the atmosphere, i.e. $T(\tau), p_e(\tau), p_g(\tau)$ etc., is obtained by solving the equation of *hydrostatic equilibrium*,

$$\frac{dp_g}{d\tau} = \frac{g}{\kappa} - \frac{\pi}{c} \frac{1}{\kappa} \int_0^\infty \kappa_\nu F_\nu d\nu \tag{1}$$

together with the condition for *flux constancy*,

$$F = \int_0^\infty F_\nu d\nu + F_{conv} = \frac{\sigma}{\pi} T_{eff}^4 = \text{independent of } \tau . \tag{2}$$

Here τ is the optical depth at an arbitrarily chosen reference frequency ν_0, κ is the corresponding absorption coefficient $\kappa \equiv \kappa_{\nu_0}$, p_g is the gas pressure, p_e the electron pressure, F_ν are the monochromatic radiative fluxes, and F_{conv} is the convective flux.

The calculation of the radiative fluxes involves solution of the equation of *radiative transfer* for the intensity I_ν (at each frequency ν) into which the absorption coefficients κ_ν enter:

$$\cos \varphi \frac{dI_\nu}{d\tau_\nu} = I_\nu - S_\nu . \tag{3}$$

For simplicity, scattering is not considered in the present discussion, and the source function S_ν is set equal to the Kirchhöff-Planck function $B_\nu(T)$.

Finally, Equation (2) has to be supplemented by the Schwarzschild criterion according to which *convective instability* occurs if

$$\left(\frac{d \log T}{d \log p_g}\right)_{\text{adiabatic}} < \left(\frac{d \log T}{d \log p_g}\right)_{\text{radiative}} \tag{4}$$

Besides the absorption coefficients κ_ν, some *thermodynamic properties* of the stellar matter have to be known in order to calculate the atmospheric structure by Equations (1) through (4).

In particular, the equation of state

$$p_g = p_g(T, p_e), \quad \rho = \rho(T, p_e) \tag{5}$$

is needed for the integration of the equation for hydrostatic equilibrium. Furthermore, the entropy is required to evaluate the adiabatic gradient and convective flux.

With the assumption of LTE, the thermodynamic quantities and the κ_ν are determined (i) by T and p_e through the Saha-Boltzmann equations, and (ii) by the abundances ϵ_i of the constituents. If molecule formation is important the equations for the molecular equilibria have to be added.

The κ_ν comprise also *line absorption* besides continuous absorption. For its calculation, in addition to the population in the lower level, the profile function has to be known. It depends among others on the partial pressures of perturbing atoms (collisional damping, proportional to p_e or p_H in most cases) and on the velocity fields (microturbulence, see Section 3.4).

It should be remarked that line absorption is not only important for determining the temperature stratification (blanketing) but has to be considered (in the ultraviolet) also in the radiative pressure term of the hydrostatic equation for early-type stars.

3. Effect of Abundances on the Atmospheric Structure

The parameters T_{eff}, g, ϵ_i which determine a model atmosphere are of different nature. While T_{eff} and g are directly contained in the basic equations, *the abundances ϵ_i enter only through the thermodynamic properties and the absorption coefficients*. An element which does not contribute to these quantities ('trace element') has no influence upon the structure of the atmosphere. However, before constituents can be neglected as trace elements in model calculations their possible influence has to be carefully examined. For example, potassium contributes about ≥ 30 per cent to the electron pressure in cool stars ($T \lesssim 3000$ K) despite its low abundance of only about 10^{-7}. ϵ_H (see, e.g. Bode, 1965).

3.1. EQUATION OF STATE

The thermodynamic quantities are sums of the contributions of the constituents so that the relative abundances ϵ_i simply enter as weights (when T and p_e are taken as independent variables). If $f_{ij}(T, p_e)$ is the fraction of element i in ionization state j, the equation of state for the case of negligible molecule formation can be written in the form (see, e.g. Mihalas, 1970):

$$p_g(T, p_e) = p_e \times \left(1 + \frac{\sum\limits_i \epsilon_i}{\sum\limits_i \epsilon_i \sum\limits_j j f_{ij}}\right) \tag{5a}$$

$$\rho(T, p_e) = \frac{p_g - p_e}{kT} \mu m_H \, ,$$

where $\mu = \sum \epsilon_i \mu_i / \sum \epsilon_i$ is the mean molecular weight (μ_i = atomic weight of element i).

For high temperatures, $p_g/p_e \approx 2$ for hydrogen-dominated matter, and ≈ 1.5 for helium-dominated matter. For lower temperatures, when hydrogen and helium are largely neutral and the metals M mostly singly ionized, $p_g/p_e \approx (\epsilon_H + \epsilon_{He}) / \epsilon_M$.

In cases of partial ionization, essentially the abundance ϵ_i competes with the factor $e^{-\chi_{ion}/kT}$ of the Saha equation (χ_{ion} ionization potential). For example, in cool stars elements with low χ_{ion} such as K become important; or in the intermediate temperature range, the CNO group significantly contributes to p_e if hydrogen is strongly deficient (e.g. in the helium star HD 30353, Nariai, 1967).

3.2. CONVECTION

Element abundances affect the Schwarzschild criterion (4) since they enter into the adiabatic as well as into the radiative gradient. For the adiabatic gradient, abundant elements and the electrons have to be taken into account according to the expressions for the specific heats or the derivatives of the entropy. The radiative gradient on the other hand is determined essentially by the mean absorption coefficient and its pressure dependence, so that the ϵ_i enter through the equation of state and the κ_ν.

An important application is the comparison of the hydrogen convection zone in cool metal-poor population II stars with that in normal stars. The lower metal abundances imply lower H^- absorption and hence, from the hydrostatic equation, a higher gas pressure at a given optical depth. This results in convection zones extending higher up into the atmospheres of metal-poor stars, and in flatter temperature gradients (Krishna Swamy, 1969).

3.3. ABSORPTION COEFFICIENTS AND TEMPERATURE STRATIFICATION

The structure of the atmosphere, essentially $T(\tau)$, depends critically on the distribution of the absorption coefficients with frequency, i.e. on the non-grayness of the κ_ν. Most weight is given to the frequency range where, roughly speaking, the maximum flux occurs. This fact can be taken into account, for example, by referring the τ-scale to the Rosseland mean opacity,

$$\frac{1}{\bar{\kappa}_R} = \int_0^\infty \frac{1}{\kappa_\nu} \frac{\partial B_\nu}{\partial T} \, d\nu \bigg/ \int_0^\infty \frac{\partial B_\nu}{\partial T} \, d\nu \, . \tag{6}$$

This is a good approximation (diffusion approximation) in the deeper layers, whereas near the surface other mean values, e.g. the Planck mean, are more appropriate.

Insight into the dependence of $T(\tau)$ upon the κ_ν, especially when line absorption is important, can be gained by simple approximations, e.g. the picket-fence model of Chandrasekhar (see Mihalas, 1970), or the approximation by Hunger and Traving (1956). The latter makes use of the moments of κ_ν, weighted by the Planck function,

$$M_s = \int_0^\infty \kappa_\nu{}^s B_\nu d\nu; \; s = -2, -1, 0, 1 \tag{7}$$

The coefficients Q, k_α and L_α in the *approximate* expression for the temperature stratification

$$T^4(\bar{\tau}_R) = \tfrac{3}{4} T_{\text{eff}}^4 \left(\bar{\tau}_R + Q - \sum_{\alpha=1}^m L_\alpha e^{-k_\alpha \bar{\tau}_R} \right) \tag{8}$$

can be characterized by only *two* moments, M_{-2} and M_1 (M_0 is essentially the normalization of B_ν, and M_{-1} is used up for the definition of the Rosseland scale). While M_{-2} is determined by the lowest κ_ν, M_1 (corresponding to the Planck mean) takes account of the peaks of the absorption, e.g. of the absorption by strong lines. The surface temperature depends essentially on M_1, approximately $T^4(o)/T^4_{\text{eff}} \propto M_1^{-\frac12}$.

Besides the distribution over frequency ('crowding of lines'), the effect of abundances on $T(\tau)$ depends also on the distribution of the line strengths on the curve of growth. For example, many overlapping faint lines or wings of strong lines will behave like a quasi-continuous absorption, and have an influence which is different from that of more isolated medium-strong lines. The general effect of *line absorption* on $T(\bar{\tau}_R)$ ('blanketing') is to increase the temperature in deeper layers ('backwarming') and to decrease it near the surface; the flux is redistributed: the decrease by the lines ('blocking') is compensated by an increase in relatively line-free ranges of the spectrum.

For a crude orientation about the order of magnitude of the contribution of various elements to the line absorption, we may make use of the fact that bound-bound and bound-free transitions steadily merge at the absorption edge, so that the height of the edge can be taken as some measure for the line absorption. In the hydrogenic approximation (principle quantum number n, effective charge Z), the absorption coefficient a_n per atom in state n is about $8 \times 10^{-18} \, nZ^{-2}$ cm^2 at the edge (see e.g. Allen, 1973). Thus we may expect line absorption of element i to be important for the structure of the atmosphere, (i) if the abundance ϵ_i is sufficiently high and the frequency ν_n of the edge falls in a range where the continuous absorption $\kappa_{\nu,c}$ is low, i.e. if $\epsilon_i f_{in} a_n^{(i)} / \Sigma \epsilon_i$ becomes comparable to $\kappa_{\nu,c}$ (f_{in}: fraction of element i in state n), and (ii) if ν_n falls in a range where the Rosseland weight function is large.

Although the inclusion of line absorption is straightforward in principle, the large number of lines which in many cases have to be taken into account, causes problems (lack of atomic data, question of completeness for lines considered, numerical treatment). Calculations with a direct inclusion of the stronger lines have, for example, been performed for the ultraviolet blanketing in early-type stars by Morton and his collaborators (e.g. Mihalas and Morton, 1965; Adams and Morton, 1968). Besides the direct approach, statistical methods have (in particular using opacity distribution functions) been developed to deal with a large number of lines (e.g. Labs, 1951; Böhm, 1954; Strom and Kurucz, 1966; Carbon, 1973; Peytremann, 1974; Gustafsson et al., 1975), or the problem of calculating line-blanketing has been avoided by using scaled (solar) temperature stratifications in a restricted range of parameters. Particularly when the influence of abundances on the model is to be investigated, an accurate and consistent treatment of the line (and associated continuous) absorption is important, a schematic incorporation of line blocking does not seem sufficient. A detailed discussion and comparison of the different methods is, for example, contained in the recent paper by Gustafsson et al. (1975).

3.4. Velocity Fields

In spectral analyses, the atmospheric velocity field is mostly idealized as being composed of microturbulence and macroturbulence. One would expect that, in principle, the velocities can be calculated from the parameters which determine the atmosphere (and the layers below it). This view is supported by observed correlations between the microturbulent velocities ξ with T_{eff}, g and the metal content (see, e.g., Baschek and Reimers, 1969; Nissen, 1970; Gustafsson and Nissen, 1972; Reimers, 1973). In practice, however, the microturbulence has to be treated as an independent additional parameter, which is derived empirically from the line spectrum, since the present theory of hydrodynamic phenomena does not allow its calculation with sufficient accuracy.

Microturbulence influences the structure of a stellar atmosphere through its contribution to the pressure and through the increase of the widths of the line absorption coefficients and hence of the line blanketing: (i) the left-hand side of Equation (1) has to be modified by adding a term $\frac{1}{2} d(\rho \xi^2)/d\tau$.* This term has to be taken into account if ξ becomes comparable to or larger than the velocity of sound, for example, in supergiants (see e.g., Groth, 1961) and possibly in some helium stars (low velocity of sound) such as HD 124448 in which Schönberner and Wolf (1974) find $\xi=10$ km s^{-1}. (ii) The Doppler width entering the line absorption coefficient is increased according to $\Delta\nu_D/\nu=(\nu^2_{thermal} + \xi^2)^{1/2}/c$. Note that the lines on the flat part of the curve of growth whose strengths are sensitive to ξ constitute an important contribution to the line blanketing in many cases.

We may expect that microturbulent velocities correlate with the velocities in the convection zones (e.g. Böhm-Vitense, 1971) which in turn depend on the element abundances. Although metal poor population II stars have hydrogen convection zones extending higher up into the atmosphere (see Section 3.2), the mixing-length theory predicts their microturbulent velocities to be smaller. This is because they are determined by non-local (overshoot) phenomena which originate from layers below the upper boundary of the convection zone (Böhm-Vitense, 1971) where the convective velocities

---

* The contribution of macroturbulence to the pressure depends on the special structure of the velocity field.

are smaller in the metal-poor stars. The observed low microturbulence in population II stars, however, is as yet not quantitatively understood (Böhm-Vitense, 1975).

3.5. EFFECTS OF ABUNDANCES IN NON-STANDARD MODELS

So far we have discussed the essential effects of abundances on the atmospheric structure with the standard assumption of a homogeneous plane-parallel and flux constant atmosphere in hydrostatic and local thermodynamic equilibrium. Non-LTE calculations, for example, in general yield level populations different from those in LTE so that quantitative differences, say in the absorption coefficients, may arise. However, as the abundances influence the structure only indirectly through the thermodynamic properties and the absorption coefficients, no *basically* different effects occur when more general assumptions are made.

Finally, it should be mentioned that some stars such as the peculiar A stars have atmospheres which are *chemically inhomogeneous*. Inhomogeneities occur across the surface, as is more or less directly observed, but possibly also abundance gradients with depth have to be considered in view of the diffusion or the magnetic accretion hypothesis for A_p stars.

In the present discussion, only the basic effects of abundances on the structure of a stellar atmosphere are considered. No attempt is made to give a systematic discussion of results obtained for various regions of the HR diagram. Recently, investigations (containing also references to previous work) of the influence of metal abundances ϵ_M on model atmospheres have been carried out by Peytremann (1974) in the range $5000 \leqslant T_{eff} \leqslant 8500$ K and $2 \leqslant \log g \leqslant 4.5$, by Böhm-Vitense (1975) for $4000 \leqslant T_{eff} < 8000$ K and $L/L_\odot \leqslant 9000 \, M/M_\odot$, and by Gustafsson *et al.* (1975) for $3750 \leqslant T_{eff} \leqslant 6000$ K and $0.75 \leqslant \log g \leqslant 3$. For the cool giants considered by these authors, differences of the metal abundance within the range $-3 \leqslant \log(\epsilon_M/\epsilon_H)/(\epsilon_M/\epsilon_H)_\odot \leqslant 0$ do not lead to very pronounced effects on the *model atmospheres*, e.g. Böhm-Vitense finds that differences of only $\lesssim 200$ K occur in the temperature stratification $T(\bar\tau_R)$. Gustafsson *et al.* find empirically that, at given $\bar\tau_R$, ΔT is proportional to $\Delta \log \epsilon_M/\epsilon_H$.

4. Theoretical and Observed Spectra

So far, we have followed the 'theoretician's point of view': a theoretical spectrum is calculated from the parameters T_{eff}, g, and ϵ_i. One can calculate grids of model atmospheres, calculate all desired spectral quantities with their dependence on the ϵ_i and hence obtain the calibrations for spectral classification. From the standpoint of applications, particularly of classification, however, the inversion of this procedure is of prime interest: observed, suitably chosen spectral criteria are to be fitted to a theoretical calibration which then allows to read off basic stellar parameters. It is desirable to know the sensitivity of the criteria to the parameters (i.e. the accuracy which can be achieved) or how far the parameters can be uniquely determined.

4.1. REMARKS ON THE BASIC STELLAR PARAMETERS

Since a (standard) stellar *atmosphere* is determined by T_{eff}, g and ϵ_i, the natural way to explore the influence of abundances on its structure and on the emergent radiation is to keep T_{eff} *and g fixed*, and to consider various ϵ_i. Comparing stars with the same T_{eff} and g is equivalent to comparing stars with the same T_{eff} and 'specific luminosity' $L/\mathcal{M}$ ($\mathcal{M}$ stellar mass).

On the other hand, the structure of the *entire star* is determined by its mass $\mathcal{M}_o$ and composition ϵ_i^o at birth, and by its age t (if rotation, magnetic fields, mass transfer from a companion etc. are neglected). Hence $\mathcal{M}$, L, T_{eff}, g, ..., *and the atmospheric composition** ϵ_i depend on $\mathcal{M}_o$, ϵ_i^o and t. (In practice, however, present theories of mass loss and mixing do not allow to calculate changes of $\mathcal{M}$ and ϵ_i during the stellar evolution). The atmospheric parameters T_{eff}, g, ϵ_i may therefore be correlated as a consequence of the star's evolution, so that it may not always be appropriate to vary the abundances independently of T_{eff} and g.

As a special example we consider stars on or near the main sequence. Here $\epsilon_i = \epsilon_i^o$ (apart from exceptional cases) because the atmospheres still have the original composition, so that $T_{eff} = T_{eff}(\mathcal{M}_o, \epsilon_i, t)$ and $g = g(\mathcal{M}_o, \epsilon_i, t)$. Conversely, if we compare two main-sequence stars with different abundances (e.g. a population II star with a population I star), but identical T_{eff} and g, their masses and ages will be different.

4.2. REMARKS ON THE UNIQUENESS OF THE DERIVED ATMOSPHERIC PARAMETERS

As spectral classification should be applicable to a large number of stars, the number of independent spectral criteria is necessarily restricted. Obviously, the number of parameters which are to be derived has to be less than or at most equal to the number of criteria. In practice, this enforces a restriction in the number of elements which can be considered individually. Usually, the elements are grouped together and abundances for the metals or for the CNO group etc. are derived upon the assumption that within each group the relative element abundances are held fixed, e.g. at their solar values.

A more subtle question concerning the uniqueness of derived parameters arises due to the fact that (as long as molecule formation is unimportant) the theoretical spectrum is determined if $T(\tau)$ and $p_e(\tau)$ are given together with the abundances ϵ_s of those elements s which exhibit *observable* structures (lines, absorption edges, ..) in the spectral range considered. This may be realized in some detail, e.g. by making use of the Eddington-Barbier approximation for the flux F_ν emerging from the stellar surface,

$$F_\nu = 2 \int_0^\infty B_\nu(\tau_\nu) E_2(\tau_\nu) d\tau_\nu \approx B_\nu(\tau_\nu = \tfrac{2}{3}),\qquad(9)$$

where τ_ν is the optical depth at the particular frequency ν under consideration, and E_2 the exponential integral of order 2. Since in spectral classification we are more concerned

---

*Note that not all kinds of observed abundance differences are directly related to stellar evolution. For example, the anomalies found in A_p stars or white dwarfs are probably due to processes which operate in the outer layers only.

with *relative* intensities, we consider the ratio of fluxes at two frequencies ν and $\nu+\Delta\nu$ with corresponding optical depths τ and $\tau+\Delta\tau$ (or absorption coefficients κ and $\kappa+\Delta\kappa$). Let τ be the reference scale for which $B_\nu(T(\tau))$ is regarded as known, then, for small $\Delta\nu$ and $\Delta\tau$,

$$\frac{F_2}{F_1} \approx 1 - \Delta\tau \left(\frac{\partial \ln B_\nu}{\partial \tau}\right)_{\tau=\frac{2}{3}} + \Delta\nu \left(\frac{\partial \ln B_\nu}{\partial \nu}\right)_{\tau=\frac{2}{3}} + \dots \tag{10a}$$

with $\Delta\tau = \int_0^{\tau=^2/_3} \Delta\kappa/\kappa \, d\tau$. For a spectral line, the variation of B_ν with frequency can be neglected over the line, and $\Delta\kappa$ is to be identified with the line absorption coefficient l, so that the depression in the line is

$$1 - \frac{F_2}{F_1} \approx \frac{2}{3} \left\langle\frac{1}{\kappa}\right\rangle \left(\frac{\partial \ln B_\nu}{\partial \tau}\right)_{\tau=\frac{2}{3}}. \tag{10b}$$

Here $\langle 1/\kappa \rangle$ denotes the ratio of line to continuous absorption coefficient averaged over the layers down to $\tau = {}^2/_3$.

We infer from Equations (10) that spectral features (such as line strengths, colours, bolometric corrections etc.) primarily depend (i) on the relative temperature gradient, and (ii) on the line and continuous absorption in the 'layers of formation'. $\langle 1/\kappa \rangle$ is (a) directly proportional to the ratio of the abundance ϵ_s of the element producing the line to that of the main contributor to the continuum (in most cases hydrogen), and depends (b) on the level populations, i.e. excitation and ionization conditions which are determined by T and p_e. The $T-p_e$ relation is therefore more directly significant for discussions of stellar spectra than the fundamental parameters themselves. Incidentally, this was pointed out already in 1958 by Unsöld who discussed the effect of different ϵ_i in cool stars, within the framework of coarse analysis, by keeping mean values of T and p_e fixed.

Is it now possible that *different* combinations of the model parameters (T_{eff}, g, ϵ_i) result in the *same* $T-p_e$ relation and in the *same* spectrum? In this generality, this seems possible only for the rather academic case that the elements s having observable spectral features (lines) are not the same elements that significantly influence the atmospheric structure. If, however, only a *selected* number of spectral criteria (e.g. only continuum and Balmer lines) and/or only a *restricted* spectral range (e.g. only the visible for early-type stars) is considered, the problem of the uniqueness is not trivial, i.e. spectral classification is more vulnerable to ambiguities than more detailed spectral analyses.

For example, in cool stars helium does not exhibit lines in the visible and does not contribute to the electron pressure and the opacity. The structure of an atmosphere with such an abundant 'passive' constituent cannot be distinguished from that of a star with e.g. no helium and a different (higher) gravity. Another important example is the role of the ultraviolet line blanketing in early-type stars: it has been shown (e.g. Mihalas and Morton, 1965) that, for the same gravity, the $T-p_e$ relation of a star without metals is very similar in the formation layers of the visible spectrum to that of a star with metals (and hence with ultraviolet line-blanketing) which has a different effective temperature.

Related problems occur, for example, for the interpretation of the spectra of A_p stars (see e.g. Leckrone, 1973), or for the connection of classification of hot stars in the extreme ultraviolet with that established in the visible.

4.3. SYSTEMATIC ERRORS IN DIFFERENTIAL ANALYSES

A theoretical calibration of classification criteria is subject to systematic errors which seem unavoidable even in sophisticated model atmosphere calculations. *Differences* in the stellar parameters ΔT_{eff}, $\Delta \log g$, $\Delta \log \epsilon_i$, however, can be obtained with higher accuracy because all kinds of sources of systematic errors largely cancel when stars of similar structure are compared. *Differential evaluation of calibrations obtained by model atmosphere techniques can be regarded as the most reliable procedure in theoretical stellar classification.* (This, of course, does not mean that absolute calibrations are dispensable).

One should realize, however, that the range of interest for element abundances is so wide that even in differential work systematic errors cannot be excluded. Consider two stars with identical T_{eff} and g, then their atmospheric structures are similar only if all relevant $\Delta \log \epsilon_i$ are small enough, i.e. if a trace element (as defined in section 3) remains a trace element, and if, for example, an important contributor to the opacity remains important, etc.

For example, in solar-type stars a decrease of the metal abundance by a factor of 10 or more leads to a reduced line blanketing and hence a different $T(\tau)$. Furthermore, strong lines in the metal-rich star which depend on the van der Waals broadening will shift onto the flat part of the curve of growth which depends on the microturbulent velocity, so that also systematic effects in the relative strengths of the lines may arise. On the other hand, an increase of the metal content in solar-type stars raises the question how far the lines included for the blanketing calculations are complete. The overabundances of iron-group elements and rare earths are found in some peculiar A stars to be so large that they dominate the line absorption in the ultraviolet and hence lead to a different $T(\tau)$ as compared to normal stars of similar T_{eff} and g (e.g. Leckrone, 1973; Leckrone *et al.*, 1974).

5. Conclusion

Quantitative stellar classification relies upon the calculation of theoretical spectra which can best be obtained by using model atmospheres. In order to minimize the systematic errors inherent in model and line computations, differential methods should be preferred.

Besides the direct ('observable') dependence of spectral features (lines, colours etc.) on the abundances of the elements which produce the respective feature, the spectrum depends also implicitly on element abundances through their influence upon the structure of the atmosphere, i.e. essentially upon the $T-p_e$ relation.

While effective temperature and gravity are directly contained in the basic equations from which the model atmosphere is to be constructed, the abundances enter only indirectly through the thermodynamic properties (mainly electron pressure and entropy) and through the absorption coefficients of the stellar matter. Since the emergent intensities depend crucially on the temperature stratification which in turn depends strongly on the absorption coefficients, the latter have to be consistent with the element abundances. In particular the inclusion of the absorption by numerous metal lines — which cannot be neglected for most types of stars — requires considerable numerical effort (even in the case of LTE). The microturbulent velocity has to be considered as an additional parameter since at present it cannot be expressed in terms of the basic parameters with sufficient accuracy.

Acknowledgment

I would like to thank M. Scholz, G. Traving and R. Wehrse for helpful comments.

References

Adams, T. F. and Morton, D. C.: 1968, *Astrophys. J.* **152**, 195.

Allen, C. W.: 1973, *Astrophysical Quantities*, The Athlone Press, London.

Baschek, B. and Reimers, D.: 1969, *Astron. Astrophys.* **2**, 240.

Bode, G.: 1965, *Veröff. Inst. Theoret. Physik u. Sternwarte*, Univ. Kiel.

Böhm, K.-H.: 1954, *Z. Astrophys.* **34**, 182.

Böhm-Vitense, E.: 1971, *Astron. Astrophys.* **14**, 390.

Böhm-Vitense, E.: 1975, in B. Baschek, W. H. Kegel, and G. Traving (eds.), *Problems in Stellar Atmospheres and Envelopes*, Springer-Verlag, Berlin, Heidelberg, New York, p. 21.

Carbon, D. F.: 1973, *Astrophys. J.* **183**, 903.

Groth, H.-G.: 1961, *Z. Astrophys.* **51**, 231.

Gustafsson, B. and Nissen, P. E.: 1972, *Astron. Astrophys.* **19**, 261.

Gustafsson, B., Bell, R. A., Eriksson, K., and Nordlund, Å.: 1975, *Astron. Astrophys.* **42**, 407.

Hunger, K. and Traving, G.: 1956, *Z. Astrophys.* **39**, 248.

Krishna Swamy, K. S.: 1969, *Astron. Astrophys.* **1**, 297.

Labs, D.: 1951, *Z. Astrophys.* **29**, 199.

Leckrone, D. S.: 1973, *Astrophys. J.* **185**, 577.

Leckrone, D. S., Fowler, J. W., and Adelman, S. J.: 1974, *Astron. Astrophys.* **32**, 237.

Mihalas, D.: 1970, *Stellar Atmospheres*, Freeman, San Francisco.

Mihalas, D. and Morton, D. C.: 1965, *Astrophys. J.* **142**, 253.

Nariai, K.: 1967, *Publ. Astron. Soc. Japan* **19**, 63.

Nissen, P. E.: 1970, *Astron. Astrophys.* **6**, 138.

Pecker, J. C.: 1973, in B. Hauck and B. E. Westerlund (eds.), 'Problems of Calibration of Absolute Magnitudes and Temperature of Stars', *IAU Symp.* **54**, p. 173.

Peytremann, E.: 1974, *Astron. Astrophys.* **33**, 203.

Reimers, D.: 1973, *Astron. Astrophys.* **24**, 79.

Schönberner, D. and Wolf, R. E. A.: 1974, *Astron. Astrophys.* **37**, 87.

Strom, S. E. and Kurucz, R. L.: 1966, *J. Quant. Spectrosc. Radiative Transfer* **6**, 591.

Unsöld, A.: 1955, *Physik der Sternatmosphären*, 2nd. ed., Springer-Verlag, Berlin, Göttingen, Heidelberg.

Unsöld, A.: 1958, *Nachr. Akad. der Wissenschaften*, Göttingen, IIa, Math.-Phys.-Chem. Abt., No.2.

DISCUSSION

R. Kandel: In connection with velocity fields, there are of course many problems. I would consider your 'second' effect, viz. the effect of line broadening, the fundamental one, since the 'velocity fields' are in fact derived from broadening observations. My remark is that, whenever the 'turbulent pressure' term in the HSE equation becomes important, it is very likely that dissipation terms also will become important. Some time ago Anthony Hearn in *A. and A.* found that the energy requirements of microturbulence were uncomfortably high in A stars. For Betelgeuse, many authors found a 'super-sonic' microturbulence of 10 km s^{-1}; if one applies Hearn's calculations to this, one finds a mechanical energy flux of the order of the total flux of the star is required by the dissipation of such micro-turbulence. Thus there is a serious problem, and velocity fields must, if they are real, be included in the energy transfer too.

Baschek: I agree, as for the fairly hot stars studied by A. G. Hearn. For cooler stars with their strong hydrogen convection zones, the problem does not seem so serious to me.

Foy: This comment concerns the turbulence. I have checked the effect in the abundance deter-minations of the changes in stellar structure due to the turbulent pressure (Foy, thesis, 1974).

Consider a model defined by T_{eff} = 4500 K, log g = 2.50 and [M/H]$_0$ = 0.0. Ignoring turbulent pressure whereas the turbulent velocity changes from 0.0 to 5.0 km s^{-1} induces errors in abundance determination less than 0.10 or 0.15 dex (depending upon the lines). So this is not very large, compared to the usual error bias (?) in such works (typically 0.20 dex).

Baschek: At T = 4500 K, the velocity of sound is about v_s = 8 km s^{-1}, so that in your example $(\xi/v_s)^2$ is at most about 0.4 and hence no drastic effects upon the structure are expected.

Bell: In the analysis of stars for which high values of ξ are found, was that high value of ξ included in the calculation of the atmosphere used for the analysis?

Baschek: In most cases, not. One should realize, however, that an inclusion of ξ in the hydrostatic equation would imply knowledge about its dependence with depth, e.g. should for some reason ξ be proportional to $\rho^{-\frac{1}{2}}$, then $\frac{1}{2}\frac{d}{d\tau}\rho\xi^2 = 0$. In special cases, e.g. for the A supergiant αCyg, some information on $\xi(\tau)$ has been derived and its effect on the pressure has been taken into account (Groth, 1961 – see references).

Schatzman: One should be very careful about assuming that the chemical composition in the atmosphere is the same as the initial composition. Time scale of diffusion, even including macroscopic transport of some sort, is sufficiently short to produce changes in chemical composition.

De La Reza: We are studying in collaboration with Dr Querci the influence of abundances on the structure on the atmosphere of carbon stars. We chose a group of elements that constitute the principal electrons donors as K, Na, Ca, Mg. The point is to study the global problem, i.e. introduce an initial abundance of the element, see what is the effect on the atmosphere, with this calculate the theoretical profile and compare with the observations. If the fit is not good we continue with this iterative scheme up to obtaining complete agreement. From preliminary results for a giant star of T_{eff} = 3500°. The effect to put 10 times more number of K and Na atoms as in the sun is to raise the temperature to 200° K in a certain region of the atmosphere. We expect that this effect is even greater in lower values of T_{eff}.

Schatzman: It is difficult to make any predictions. For stars earlier than late F for example, the diffusion takes place mainly at the bottom of the convective zone. The velocity of diffusion varies very much with the depth of the lower boundary of the HCZ. We can expect age effects depending upon the spectral type. One more difficulty is due to the inhibition of diffusion under gravitational (or radiation pressure) drag by macroscopic motions (stochastic motions). Anyhow, differences are to be expected when comparing main-sequence stars of different age and of spectral type earlier than F8 for example.

Garrison: I would like to comment on a small, but extremely important misconception about spectral classification which stellar atmospheres theoreticians often express. While it is true that MK spectral classification uses a restricted number of line ratios as *Guidelines*, the ultimate criterion is that the entire blue-violet spectrum matches that of the standards. This considerably reduces the chance for error because it is very rare that in the entire spectrum, abundances will mimic T_{eff} and log g. Thus the star will be called peculiar by the perceptive classifier, even though he may not know exactly *why* it doesn't match. If something is wrong, it usually doesn't match in detail and the classification is a signal to the high-dispersion spectroscopist of theoretician to investigate the star in more detail to determine why.

Baschek: I think there is a steady transition from classification based on very few parameters (e.g. two or three colours) over spectral classification to detailed model analyses. In my talk, I used the term 'spectral classification' or 'classification' in a general sense.

Kandel: Are electrons still important in cool stars? If so, as you pointed out, potassium is not a trace element, but then one has to worry about non-LTE effects. Auman and Woodrow in a recent *Astrophysical Journal* have shown departures from the Saha equation for cool giants and supergiants. The critical thing appears to be the collision of cross-sections with neutral atoms and molecules.

Baschek: At temperatures of about 3000 K electrons are still important. Regarding non-LTE effects, I agree that they have to be carefully examined whenever some indications for their importance show up.

Jaschek: You have mentioned some work relating to stellar atmosphere calculations done for different metal abundances. My question is if you could detail a little bit more which are the regions of the HR diagram covered by such studies.

Baschek: To my knowledge, systematic abundance studies are more or less restricted to the region (F-K) where the 'metals' as a group are important for the atmospheric structure. In early-type stars and in cooler stars, where molecules are important, investigations are complicated by the fact that more than one abundance has to be considered to vary.

Bidelman: Could you expand your remark concerning spectral classification in the far ultraviolet a bit?

Baschek: In early-type stars, the bulk of radiation is in the far ultraviolet. I wonder how far a classification system which has been introduced outside this range, i.e. in the visible, can be extended into the ultraviolet or how far it will match an ultraviolet classification scheme. Will it be too coarse, will ambiguities arise or ...?

De La Reza: As far as the influence of the changes in the electron density is concerned we have to consider not only the abundance effects but also the non-LTE effects. For instance, Auman and his collaborators are investigating the non-LTE effects for M stars and these are quite large. In the future we have also to investigate the abundance effect on this.

SYNTHESIS OF CLASSIFICATION
DISPERSION SPECTRA FROM STELLAR
ATMOSPHERES THEORY AND FROM HIGH DISPERSION SPECTRA

A. W. IRWIN, C. T. BOLTON and R. F. GARRISON

David Dunlap Observatory, University of Toronto

Abstract. The ATLAS programme has been corrected and modified for use at low effective temperatures. A grid of unblanketed model atmospheres has been generated for the region G5-K5, V-II. A spectrum synthesis programme has been written for the calculation of flux vs wavelength for selected regions of the spectrum. Temperature distributions from the grid of continuum models and from published blanketed models will be used along with published oscillator strengths as input data for the spectral synthesis programme.

Spectrograms of MK standards have been taken at 12 Å mm^{-1} and 120 Å mm^{-1}. The University of Toronto PDS microdensitometer system has been programmed to allow comparison between the synthesized spectra and the observed spectra to determine the effects of effective temperature, surface gravity, microturbulence, and abundance on classification criteria.

DISCUSSION

McCarthy: This is a most encouraging union of mathematical synthesis and observational spectroscopy.

What happens in the construction of the synthetic spectra for lines which lack determinations of oscillator strength. Is this line omitted or is some likely value assumed?

Garrison: We make the best guess we can. We don't expect the results to be unambiguous, but it is worthwhile to do the best we can and to try to interpret the results. We hope to learn something in the process.

Stephenson: For years I have been trying to learn if anyone knows the reason for the well-known depression of the continuum on the blue side of the G-band. I suspect this might be related to the violet opacity source in C and S stars, which sets in at the same position approximately. Does the fact that your synthetic spectra do not show the G-band break mean that its cause is indeed unknown, or that you have omitted something known?

Garrison: The synthetic spectra shown were actual 12 Å mm^{-1} spectrograms convolved with a 2 Å slit to simulate 120 Å mm^{-1} spectra. We have yet to reduce the 120 Å mm^{-1} plates but the comparison between the 12 Å mm^{-1} reduced spectra, the synthesized atmospheres and the 120 Å mm^{-1} spectra should show the same results. If they don't, we'll try to figure out why they don't.

Stephenson: Perhaps this means that it is difficult to use high-dispersion spectra to interpret low-dispersion ones.

Bell: The answer to Stephenson's question is, I think, that there are far fewer CH lines on the red side of the G-band compared to the blue side.

Houziaux: (1) What type of instrumental profile are you taking? It may depend pretty much on the instrument which has been used. For instance if it is a prismatic instrument you have to change the instrumental profile throughout the spectrum.

(2) What is general philosophy of such a work? Is it to determine the spectral characteristics of the MK types with the help of synthetic spectra? Why not compare directly synthetic spectra to observed spectra and forget about MK classification? There are far more parameters in synthetic spectra than in the spectral classification.

Garrison: (1) The spectrographs used are all grating instruments and the slit has been set so that

B. Hauck and P. C. Keenan (eds.), Abundance Effects in Classification, 17–18. *All Rights Reserved.*

the projected slit is slightly larger than the resolution of the plate. Irwin has not yet decided on a final profile to be used and it can be changed easily.

(2) We are comparing directly the synthesized classification dispersion spectra from high dispersion and from stellar atmospheres models with actual low dispersion spectra in order to understand some of the effects of changes in physical parameters (such as abundance, microturbulence) on classification dispersion spectra.

Bell: Why do you not try the instrumental profile $e^{-|\Delta\lambda/q|}$ advocated by C. Veth (the last BAN) instead of a Gaussian?

Secondly, I am worried that you can never see the continuum in the G-band region of your stars and this may mean that you can never convolve a 12 Å mm^{-1} spectrum to a 120 Å mm^{-1} one.

Garrison: Irwin is worrying about the profile and can easily change it, but has not yet decided which to use. I am sure he is aware of the work of Veth, but I will convey your comments to him.

The rectified tracings were obtained by fitting a continuum point at each end of the spectrum. We do not see a continuum point within the G-band even at 12 Å mm^{-1}.

A MODEL-ATMOSPHERE ANALYSIS OF
THE SPECTRUM OF ARCTURUS

R. MÄCKLE and H. HOLWEGER

Kiel University, G. F. R.

and

R. and R. GRIFFIN

Cambridge University, U.K.

Abstract. We have analysed the spectrum of Arcturus (K2 III) relatively to the Sun, using a differential technique employing empirical models for both stars. We derive an effective temperature of 4260 ± 50K and a surface gravity log g = +0.90 ±0.35; these in turn lead to a very low mass, in the range 0.1 to 0.6 $M_\odot$. Elements are found to be underabundant by an average factor of 4 compared with the Sun. The abundance patterns in the two stars are significantly different, in keeping with the belief that Arcturus is a star of an older generation than the Sun. The carbon isotope ratio, which is as small as 5 or 6, shows that the atmospheric material of Arcturus has been processed through the CNO cycle, and theoretical arguments also indicate that Arcturus is somewhat evolved.

Reference

Mäckle, R., Holweger, H., and Griffin, R. and R.: 1975, *Astron. Astrophys.* 38, 239.

DISCUSSION

Walborn: If the '*p*' were dropped from the spectral type, then Mme Cayrel would complain about the classification! *Some* designation must be attached to indicate that the spectrum shows a significant difference from that of the standard star.

Keenan: The designation *p* can be dropped from the classification if we add the designation CN-1 to indicate that the star is moderately deficient in CN.

Bidelman: What deficiencies did you get for C, N, and O? This is useful in judging how deficient in C and N a star must be to be termed CN-weak on classification plates, for Arcturus is a relatively mild case of CN deficiency.

Griffin: Carbon, nitrogen and oxygen were found to be deficient by 0.7, 0.9 and 0.6 in the logarithm, respectively, relative to the Sun.

Nissen: You mentioned that you have used empirical models for the Sun and Arcturus. Does that mean that you have used a scaled solar $T(\tau)$-relation in the model of Arcturus?

Griffin: Yes – to construct a preliminary model. The model underwent considerable modification and refinement in the light of various published photometries, etc.

Osborn: I am interested in the gravity you determined. I believe the preliminary curve-of-growth analysis gave log g of about 1.7 and the model atmosphere analysis gave a value about 0.9. Is this a typical error range for the determination of gravities from high dispersion spectra?

Griffin: Gravities determined from curve-of-growth analysis are inaccurate parameters; I have no hesitation in preferring the result for log g given by the empirical model analysis.

B. Hauck and P. C. Keenan (eds.), Abundance Effects in Classification, 19–20. All Rights Reserved.

Spinrad: Can you pinpoint the difference in metal abundances between the present determination (about −0.6) and the older curve-of-growth determination?

Griffin: It is difficult to make suitable allowance for the varying degrees of ionization in a curve-of-growth analysis, and in Arcturus elements such as Fe are only partially ionized in the line-forming region. Their treatment *should* be more reliably handled by a stratified model analysis.

Spinrad: The derived (low) spectroscopic mass of α Boo is very exciting, puzzling and important. It will have an impact on mass loss and white dwarf mass and radius computations.

INFLUENCE OF UNCERTAINTIES OF MOLECULAR DATA UPON THE DETERMINATION OF ABUNDANCES IN COOL STARS

A. J. SAUVAL

Observatoire Royal de Belgique, Uccle, Belgium

Abstract. In cool stars, the determination of abundances of elements requires the resolution of the chemical equilibrium for each element even if atomic lines only are used. In the past, several authors have solved the well-known system of equations for a limited number of elements and of molecules. New calculations have been performed with special emphasis on the completeness of the system. Thus, all stable elements (83) have been included and a lot of molecules have been selected in order to omit no abundant compound. Owing to the lack of data (since many molecules are as yet not known spectroscopically), it appeared necessary to estimate many molecular parameters. Approximate equilibrium constants have been determined from analogies found among known molecules.

It has been shown once more that the dissociation energy is by far the most important parameter, which yields alone the final accuracy.

Our calculations include about 1200 molecules, of which more than half are new compounds never introduced in earlier investigations. We have found that, for about twenty elements such as Ti, Zr, La and most of the lanthanides, the new molecules play an important role in the chemical equilibrium. The cases of *titanium, zirconium* and *lanthanum* have been particularly investigated owing to the use of several bands of their oxides in the spectral classification of M- and S-type stars. We have noted that the experimental determinations of the dissociation energy of most monoxides and dioxides are as yet rather inaccurate. Furthermore, neither the spectrum nor the heat of atomization of hydroxides are known in the laboratory. It has appeared that estimating the dissociation energy of hydroxides was particularly difficult in consequence of the lack of data for such molecular compounds.

For oxygen-rich stars, we have investigated the influence of uncertainties of the dissociation energy of the most abundant species on the march of the molecular concentrations. In the case of *titanium*, we have found that the concentration of TiO strongly changes at low temperature ($T < 2520$ K) according to the adopted D_0°- values for TiO, TiOH and TiO_2. Therefore, the interpretation of the TiO bands in the coolest stars has to take this new effect into account. In any way, we have checked that the TiO concentration always increases from early K- to late M-type stars. We have also shown that the depletion factor for Ti presents a final uncertainty of a factor of about ten at 2520 K, which still increases at lower temperature. That is due to the cumulative effect of the inaccuracies of molecular data. Therefore, in late M-type stars, the determination of the abundance of titanium remains rather uncertain even if it is derived from Ti I or Ti II lines. The present inaccuracy will only be reduced from very precise measurements of the dissociation energy of the relevant molecules. For Zr and La, we have found similar results to that for Ti. Furthermore, we have noted that there is no determination of absolute transition probabilities for the ZrO and LaO bands (contrary to TiO bands) which are both of first importance for the spectral classification of cool stars.

For carbon-rich stars as well as for S-type stars (i.e. O/C very near unity), we have checked that the molecules always play a much less important role that in oxygen-rich stars, as it was first noted by Tsuji (*Astron. Astrophys.* **23**, 411, 1973).

Our conclusion is that there is an urgent need of accurate measurements of the dissociation energy of many molecules and also of band oscillator strengths for most of the transitions of stellar interest. Furthermore, several new selected compounds have to be investigated in the laboratory, especially the dioxides and hydroxides. Such accurate molecular data are necessary in order to get a better knowledge of the physical conditions in cool stellar atmospheres.

The detailed account of the present paper will be shortly submitted to *Astronomy and Astrophysics*.

B. Hauck and P. C. Keenan (eds.), Abundance Effects in Classification, 21–22. All Rights Reserved.
Copyright © 1976 by the IAU.

A. J. SAUVAL

DISCUSSION

Bidelman: Do you feel that there is some question about the identification of CaOH in M dwarfs? I believe that there is excellent evidence for this polyatomic molecule.

Sauval: The identification of CaOH in dwarf M stars (and not in M III stars) has been proposed by Dr. Pesch (*Astrophys. J.*) and is only based on the criterium of wavelengths coincidence. For me, such a proposed identification has to be checked by other criteria. Moreover, the laboratory analysis of the spectrum of CaOH is rather poor (see Herzberg, 1967: Spectra of polyatomic molecules). Nevertheless, Tsuji's (1973) calculations and my own results show that CaOH is the dominant molecule for $\theta > 2$, adopting D_0° (Ca-O-H) = 8.9 eV. All these results make the identification of CaOH probable, but not definitely sure.

ON THE SPECTRAL CLASSIFICATION OF COOL WHITE DWARFS

I. BUES

I. Mathematisches Institut der Freien Universität Berlin

Abstract. The Eggen-Greenstein (1965) classification of cool white dwarfs (DC, DF, DG, λ 4670, λ 4135) was based on the features visible in the spectra of low dispersion. Detailed model atmosphere computations ($11000 \geqslant T_{eff} \geqslant 7000$ K) and comparison with spectra observed by Wegner (1974, 1975) show that the variety of the spectra is due to changes in the abundance ratios of C/O, C/H and H/He only. Thus the relative strengths of the C_2 and CH-bands can be used for classification and guess of T_{eff} in the range of $T_{eff} \geqslant 6000$ K.

According to the characteristics of white dwarf spectral types reported by Greenstein (1960) the white dwarfs were classified by comparison of the strongest features with MK spectral type criteria for the main sequence, if possible (DA H I, DO He II, He I, H I, DB He I, DF Ca II, DG Ca II, Fe I, DK Ca I). Stars not fitting in this scheme were named after the strongest features visible in their spectra (λ 4135, λ 4670, λ 4670p, C_2) or DC, if no lines deeper than 10% could be detected. If this variety of spectra is due to large differences in abundances it can be determined by model atmospheres only.

For effective temperatures > 12 000 K computations by Strittmatter and Wickramasinghe (1971), Shipman (1972) and Bues (1970) showed that the classification parameter is the ratio of H/He, being >100 for spectral type DA and $<10^{-4}$ for spectral type DB. For cooler stars this ratio cannot be derived directly due to the high excitation potential of He I lines. From the high pressure necessary for the line profiles of Ca II H and K, Wegner (1972) found agreement with H/He$\leqslant 10^{-4}$ for 3 DF and DG stars but different abundances for the heavy elements. Detailed investigations of ionization-dissociation equilibria of carbon molecules in helium-rich model atmospheres (Bues, 1973) showed that a second parameter, the ratio of C/O, is necessary for the interpretation of the spectra. Stars belonging to the spectral type λ 4670 can change this ratio from 3 to 13. New observations by Wegner (1974, 1975) revealed some stars with weak H I lines and some with weak CH and C_2 bands in the range of DC stars. Our analysis (Bues and Wegner, 1975) of 4 stars yields a ratio of C/H ranging from 3 to $\frac{1}{3}$ for a helium-rich composition (H/He $\leqslant 10^{-3}$). The four parameters, the abundance ratios of H/He, He/C, C/O and C/H, are responsible for the structure of the atmospheres and the different strengths of the bands. They vary among white dwarfs of the same spectral type but differing T_{eff}, the most likely reason being a gravitational sorting mechanism.

References

Bues, I.: 1970, *Astron. Astrophys.* **7**, 91.

Bues, I.: 1973, *Astron. Astrophys.* **28**, 181.
Bues, I. and Wegner, G.: 1975, *Astron. Astrophys.* (in press).
Greenstein, J. L.: 1960, *Stars and Stellar Systems* 6, Univ. Chicago Press.
Shipman, H. L.: 1972, *Astrophys. J.* 177, 723.
Strittmatter, P. A. and Wickramasinghe, D. T.: 1971, *Monthly Notices Roy. Astron. Soc.* **152**, 47.
Wegner, G.: 1972, *Astrophys. J.* 172, 451.
Wegner, G.: 1975, *Monthly Notices Roy. Astron. Soc.* 171, 637.
Wegner, G.: 1975, *Monthly Notices Roy. Astron. Soc.* 171, 529.

DISCUSSION

Bell: Do you have an estimate of the C/N ratio in the λ 4670 white dwarfs?

Bues: From the absence of the cyanogen bands a ratio of C/N > 10 has to be assumed for $T_{eff} \geqslant$ 8500 K. In the lower temperature range C/N $\geqslant$ 100 is necessary.

Bell: Would you comment on the identification of the λ 4135 feature with bands of the He_2 molecule?

Bues: As already pointed out by Bues (1973) the interpretation of the λ 4135 feature in EG 129 as He_2 cannot be established by model atmosphere calculations. In the range of T_{eff} possible for a reasonable amount of N_{He_2} in the excited states a log $gf \approx$ 1.0 would have to be assumed for the visibility of the 5δ transition. That value is unreasonably large and, in addition, would result in a broad feature from 4100–4200 Å and not in a sharp line as indicated from the reproduced spectra and polarisation measurements. In the light of the atmospheres with a different carbon abundance which account not only for the λ 4670 stars but also for some of the DC stars, an identification of the peculiar feature with the C II line λ 4267Å seems to be more likely. Tentative model atmospheres with magnetic field strengths of 10^6-10^8 G show that an abundance ratio of He/C $\approx$ 50 and B $\approx$ 10^7 G at T_{eff} = 12000K, log g = 8 can produce a feature at the position of the band and with the observed equivalent width.

EFFECTS OF ABUNDANCE CHANGES ON THE SPECTRA OF VERY COOL WHITE DWARFS

R. WEHRSE

Lehrstuhl für Theoretische Astrophysik, Universität Heidelberg F.R.G.

Abstract. In order to study the atmospheres of cool white dwarfs of spectral types DA, DF and DG, model atmospheres have been calculated for temperatures $5000 \leqslant T_{eff} \leqslant 7000$ K and gravities $7 \leqslant \log g \leqslant 8.5$. The helium to hydrogen abundance ratio ranges from 0.1 to 1000. The abundances of the metals ϵ_M (relative to hydrogen) vary between the solar value $\epsilon_M^{\odot}$ and $\epsilon_M^{\odot}/1000$. The models are hydrogen and metal line blanketed and in radiative equilibrium. The models with increased He abundance are characterized by a decreased mean opacity since the absorption cross-section of He is much smaller than that of H^-. As a result, the spectral lines become stronger and – due to the much higher pressures – much broader. A reduction of the metal contents also increases the pressures. In this case it is caused by the shortage of electron suppliers. Here the spectral lines also become broader, but in general weaker.

These effects strengthen the overlap of the lines due to the high gravity of white dwarfs and more or less lead to a quasi-continuum which may be much below the true continuum. Because of the wavelength dependence of the line density the depression is strongest in the UV and blue spectral region, whereas for wavelengths $\lambda \geqslant 4500$ Å the difference from the true continuum is only a few per cent. The details will be published elsewhere.

DISCUSSION

Kandel: In evaluating the electron depletion by formation of negative ions, did you take into account the pressure detachment of these negative ions?

Wehrse: Pressure ionization has been taken into account by means of a very simple hydrogenic approximation, i.e. in the case of negative ions it only enters via the cut-off of the partition function of the neutral species.

Querci: How many molecules have you introduced in your calculations?

Wehrse: H_2, H_2O, CH, C_2, CN, CO, CO_2, NH, NO, N_2, OH, MgH, CaH, and TiO have been taken into account.

PART II

SOME COMMENTS ON A CATALOGUE OF
ATMOSPHERIC PARAMETERS AND
[Fe/H] DETERMINATIONS

SOME COMMENTS ON A CATALOGUE OF ATMOSPHERIC PARAMETERS AND [Fe/H] DETERMINATIONS*

G. CAYREL DE STROBEL

Observatoire de Meudon, Paris

Abstract. A few examples are given showing the utility of the catalogue of iron/hydrogen determinations for astrophysical researches.

These are:

(a) Histograms of dwarfs and giants analyzed in detail with spectral type between O and M.

(b) Logarithmic abundances diagrams of given stars.

(c) The impact on spectral classification of high-dispersion-detailed-analysis results.

(d) The use of well-determined chemical and physical parameters of nearby stars for the construction of HR diagrams in the $(M_{bol} - \log T_{eff})$ plane.

The spectra of F, G and K stars having the iron lines at their best visibility, special emphasis has been given to the results concerning these stars.

1. Introduction

The need for a catalogue of metal abundances of spectroscopically analyzed stars is felt more and more. Our actual knowledge of the chemical history of the Galaxy is based upon such metal/hydrogen determinations.

The discrepancies between different metal/hydrogen determinations in stellar atmospheres, whether based on various photometric systems or on spectroscopic analyses have put the individual astronomer in a very uncomfortable position of doubt faced with these different abundance results.

A list of metal/hydrogen values based on fairly homogeneous observing material (dispersion better than 20 Å mm^{-1}) and using a homogeneous model atmosphere technique could become a working tool for astronomers interested in abundance problems and abundance dilemmas. We present this catalogue not only for the use of such astronomers but also for others who simply need a summary of chemical composition of stellar atmospheres. In assembling this catalogue, we hope to help astronomers concerned by the impact of abundances in spectral classification, photometric systems, the study of: internal structure, evolutionary tracks and isochrones, galactic structure, heavy metal enrichment in the Universe, and last but not least, to advance our understanding of the early history of the Galaxy.

In carrying out this task, we have been faced with the following problems:

(1) How many stars have been subjected to detailed analyses?
(2) Are analyses of 'normal' stars available everywhere in the HR diagram?
(3) What are the percentages of stars analyzed in the different parts of the HR diagram?
(4) How many times has a given star been analyzed in these last twenty years?

* See appendix, pp. 223—259.

B. Hauck and P. C. Keenan (eds.), Abundance Effects in Classification, 29—46. All Rights Reserved.
Copyright © 1976 by the IAU.

(5) How much does the metal hydrogen ratio for a given star differ between different authors?

(6) What can we say about the metal/hydrogen ratio of the Hyades dwarfs in comparison to the same ratio in the Sun?

(7) What can we say about abundance peculiarities in metal-deficient stars, barium stars etc...?

(8) To what extent can a detailed analysis change the meaning of the spectral type of a given star?

These are some of the questions the authors wanted to answer in preparing the catalogue.

2. Description of the Catalogue

We have taken as metal/hydrogen parameter in the stars contained in the catalogue the logarithmic difference between the relative abundance of iron in a given star and the relative abundance of iron in a standard star. This difference is written in the form: $[Fe/H]^{star}_{stand}$. The $[Fe/H]^{star}_{stand}$ in the next to last column comes exclusively from detailed analyses based on high dispersion spectra.

The question arises when starting the catalogue, what other kinds of parameters could be useful to know at once with the iron/hydrogen abundance in a star. There were of course the atmospheric parameters with which the analysis giving the reported [Fe/H] ratio had been done: effective temperature, gravity and microturbulence, or if the analysis was a coarse analysis the excitation and ionization temperatures and the electron pressure. As other useful parameters the authors have chosen: the apparent and absolute magnitude; the spectral type in the MK system; and some photometric data. Kinematic data have been omitted.

The catalogue is composed of two tables. In Table I the authors have found it useful to report the absolute abundance determinations of iron in the solar photosphere as found by various authors. These values have been taken from Blackwell (1974) and are given on the scale $\log N_H = 12.00$. In Table II the iron/hydrogen abundances are given of about 500 stars. The description of this table is given in the introduction to the catalogue and is omitted here.

3. Methods of Abundance Analysis

In the background of each abundance determination from detailed analysis there are two kinds of astronomical 'Workers': the first are interested in the structure of the atmosphere of stars, via the theory of line formation. They calculate model atmospheres and try carefully to improve the physics of the models. The second are more interested in the kinds of stars for which they want to know the atmospheric parameters and the abundances, and they apply the model-atmosphere technique to get detailed abundances – and are not so interested in the physics of the atmosphere of the star itself. Roughly the first group belongs to Com. 36 and the second to Com. 29.

The 'first ones' are sometimes anxious about the physics involved in the models. The 'second ones' are sometimes nervous about the doubts of the 'first ones'.

To avoid the uncertainties of oscillator strengths Greenstein introduced in the late fifties the technique of the differential curve of growth analysis. The principle is to obtain the abundances of the elements in a star A relative to the abundances in a star B, taken as standard. If both stars have nearly the same spectral type, the same spectral lines can be used in both objects and the knowledge of the oscillator strengths is no longer needed. Another advantage of this method is that it cancels systematic errors in equivalent widths measurements if the same spectroscopic equipment is used to get the spectra of both stars. A third advantage of the differential method is that departures from local thermodynamical equilibrium (LTE) are eliminated if they are expected to be about the same in both stars.

The [Fe/H] results contained in the catalogue come mostly from differential detailed analyses. These results are scattered over the last twenty years. In the beginning, say until 1965, the analyses relied upon the one-layer approximation but gradually as model computations became routine, thanks to some of the great producers of model-atmospheres (like Mihalas, Harvard-Smithsonian and the school of Kiel), and to the big computers, the analyses relied upon appropriate model atmosphere computations.

The recent early spectral type [Fe/H] results rely chiefly upon the Mihalas (1966) grid of model atmospheres and of those of Kurucz et al. (1973). The recent late-type [Fe/H] results rely upon the models of Carbon and Gingerich (1969), Peytremann (1974), Böhm-Vitense (1975), Mäckle et al. (1975) and Gustafsson (1975).

A critical discussion of the abundance results of the effective temperatures and gravities will be given in the next section. Special emphasis will be given to the results concerning F, G and K stars.

4. Critical Discussion of the Temperature Gravity and Abundance Results

As we are first presenting the catalogue to this Symposium, which is devoted to abundance effects on spectral classification, the first critical examination will be made of stars which are evidently misclassified. But let us proceed according to the order of questions asked in the Introduction:

4.1. NUMBER OF STARS SUBMITTED TO DETAILED ANALYSIS

The authors have been surprised by the great number of [Fe/H] determinations, almost 500 stars with known metal/hydrogen content could become a very comfortable sample for venturing some statistical conclusions on the metal/hydrogen content in the Galaxy.

4.2. ARE ANALYSES OF NORMAL STARS AVAILABLE EVERYWHERE IN THE HR DIA-
     GRAM?

Yes, if we judge as normal the result $[Fe/H]^{star}_{stand} = 0$, the standard having a normal solar chemical composition.

4.3. WHAT ARE THE PERCENTAGES OF STARS ANALYZED IN THE DIFFERENT PARTS OF
     THE HR DIAGRAM?

The number of stars of a given spectral and luminosity class having $[Fe/H]$ determinations changes very much from one end to the other of the HR diagram.

The following histograms give a concrete view of these numbers: the histogram on Figure 1 deals with dwarfs and subgiants, that of Figure 2 with III and II luminosity class giants.

Figures 1 and 2 show that there are very few $[Fe/H]$ for O and B stars, and almost none for M and later type stars. This is obvious because the Fe lines are not at their best visibility in the spectre of these kinds of stars.

In the interval of A, F, G and K stars, there are some very high peaks on the two histograms.

On Figure 1 the A_m and A_p stars peak can be interpreted as the need to have a great

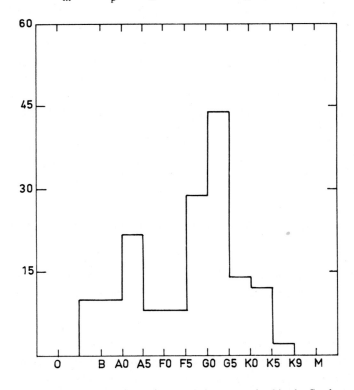

Fig. 1.   Histogram of dwarfs and subgiants contained in the Catalogue.

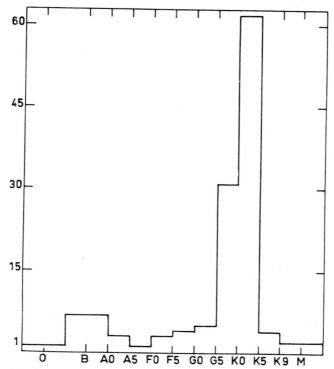

Fig. 2.   Histogram of giants of luminosity classes III and II contained in
the Catalogue.

sample of such stars to facilitate the study of their peculiarities and to find out the cause
of their peculiarities.

The peak of F and G dwarfs and subgiants is due to the need to improve our actual
knowledge of a galactic structure.

On Figure 2 the peak of G and K giants proves the interest of astronomers in studying
population effects and also evolution problems as the existence of the asymptotic giant
branch.

Questions from 4 to 7 relate to the importance of good metal/hydrogen determinations and the use that can be made of them in several astrophysical problems. They will
be answered by means of abundance diagrams. These diagrams are now very familiar to
'abundance people'; they show the values of the logarithmic metal/hydrogen or metal/-
iron differences between the star and the standard star in respect to all the elements,
analyzed in the atmosphere of a star. As an example, we present such diagrams for: the
star: 85 Peg (Figure 3), HR 3018 (Figure 4), the three Hyades dwarfs, N 63, 64, 73
(Figure 5), the four Hyades giants: γ Tau, δ Tau ϵ Tau, θ^1 Tau (Figure 6).

Furthermore we present in Figure 7 an $\left[\dfrac{N_{el}}{N_H}\right]_\odot^*$ versus the elements analyzed in the

star, of the most metal-deficient giants (more than by a factor of 5) contained in the

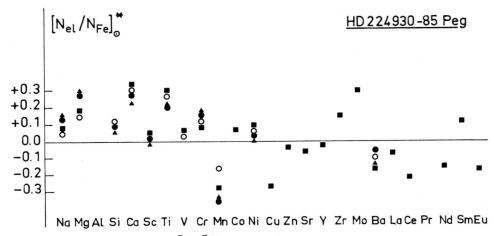

Fig. 3. Logarithmic abundances $\left[\dfrac{N_{el}}{N_{Fe}}\right]^{*}_{\odot}$ of the elements measured in 85 Peg by Wallerstein 1959 (filled circles), Wallerstein 1961 (triangles), Helfer 1963 (squares), Spite 1968 (open circles). This diagram illustrates the relative agreement between different detailed abundances analyses.

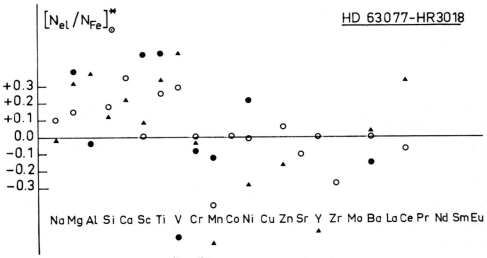

Fig. 4. Logarithmic abundances $\left[\dfrac{N_{el}}{N_{Fe}}\right]^{*}_{\odot}$ of the elements measured in HR 3018 by Kondo 1957 (filled circles), Hearnshaw 1972 (triangles), Da Silva 1975 (open circles).

catalogue, and we compare this diagram to that of the very metal-deficient subdwarfs, and to that of the barium stars of the list of Warner (1965), Figures 8 and 9. We leave to the people interested in nucleosynthesis or diffusion problems to interpret these diagrams, without, however, emphasizing that the dwarfs and the giants of the Hyades seem to have a normal solar iron/hydrogen abundance.

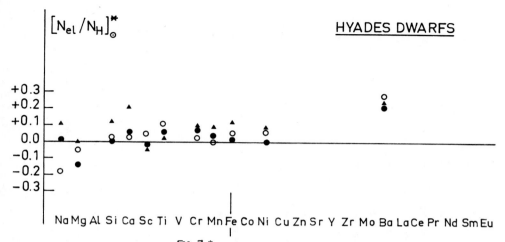

Fig. 5.   Logarithmic abundances $\left[\dfrac{N_{el}}{N_H}\right]^*_\odot$ of the elements measured in three dwarfs of the Hyades: N 63 (filled circles), N 64 (triangles), N 73 (open circles). The abundance differences in the three dwarfs do not exceed the given error bar: ± 0.2 dex.

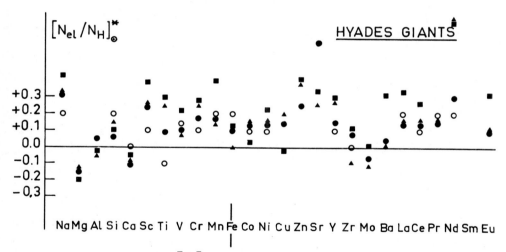

Fig. 6.   Logarithmic abundances $\left[\dfrac{N_{el}}{N_H}\right]^*_\odot$ of the elements measured in the four giants of the Hyades: γ Tau (filled circles), δ Tau (triangles), ϵ Tau (squares), θ^1 Tau (open circles).

Question 8 is the most pertinent and the most critical: 'TO WHAT EXTENT CAN A DETAILED ANALYSIS CHANGE THE MEANING OF THE SPECTRAL TYPE OF A GIVEN STAR?'

The spectral classification is basically a two-dimensional classification and it is not surprising that its interpretation becomes a problem when a third parameter (metal/hydrogen) is varied. Therefore one should avoid translating the spectral type of highly

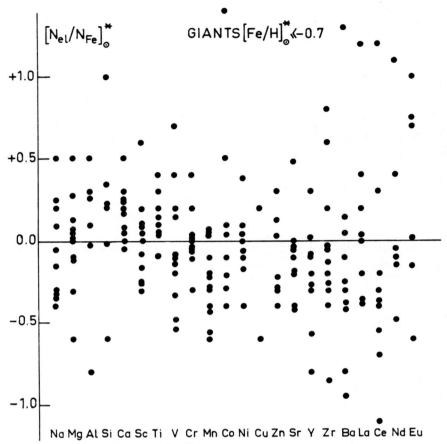

Fig. 7.   Logarithmic abundances $\left[\dfrac{N_{el}}{N_{Fe}}\right]^{*}_{\odot}$ of the elements measured in the most metal-deficient giants contained in the Catalogue.

metal-deficient stars into effective temperature and luminosity as currently done for population I stars.

As an example, let us take HD 122563, HD 165195, HD 221170 the most metal-deficient stars known until now. In our catalogue HD 122563 is classified as G0IV and has an iron/hydrogen abundance of [Fe/H] = −2.70. It was submitted to several analyses which gave as result that its effective temperature and luminosity correspond actually to a spectral type K2II of a population I star.

In my view, this is one of the major problems in spectral classification, because I would never have chosen HD 122563 for an observing programme of halo-population bright giants knowing only its spectral type.

The Table I contains other such extreme examples of discordance between MK classification and detailed analysis classification.

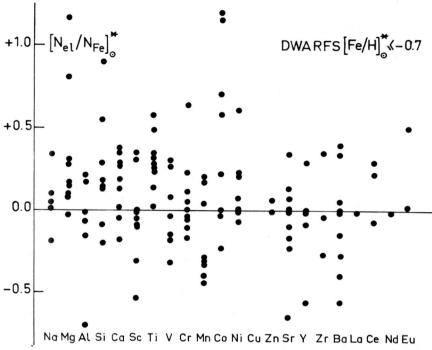

Fig. 8.   Logarithmic abundances $\left[\dfrac{N_{el}}{N_{Fe}}\right]^{*}_{\odot}$ of the elements measured in the most metal-deficient dwarfs contained in the Catalogue.

The columns of Table I are self-explanatory except for column 5, which contains the space motion vector V in the direction of galactic rotation.

The stars in Table I are in order of decreasing metal/hydrogen values. They are all high-velocity stars, and the detailed analysis performed on them has changed for all of them either their spectral types or their luminosity class, or both spectral type and luminosity.

In Table I by spectral type 'from detailed analysis' we mean the spectral type of a population I star having the same effective temperature and the same gravity as the analyzed star.

Using the data of Table I a diagram has been drawn: luminosity class vs spectral type (Figure 10). In the diagram, three kinds of symbols have been employed: open circles stand for metal-poor stars by factors ranging from 100 to 500 with respect to the Sun; crosses stand for metal-poor stars by factors from 10 to 60; filled circles, for metal-poor stars with factors from 7 to 3. On this diagram the arrows connect the two classifications (columns 4 and 5 of Table I) given for each star. The diagram shows clearly that the most metal-deficient stars have been the most shifted.

The fact that these stars have been misclassified is easy to understand. In the MK classification the spectra have been inspected in the blue-region which shows very weak lines. These stars have therefore seemed to be of an earlier spectral type than they are in reality. The most deficient halo-giants have also been misclassified in luminosity. It is well known, that for the same spectral type, but not luminosity class, the spectra of highly luminous stars have more enhanced ionized lines than the spectra of the corresponding dwarfs. Therefore the astronomers responsible for the MK classification were completely

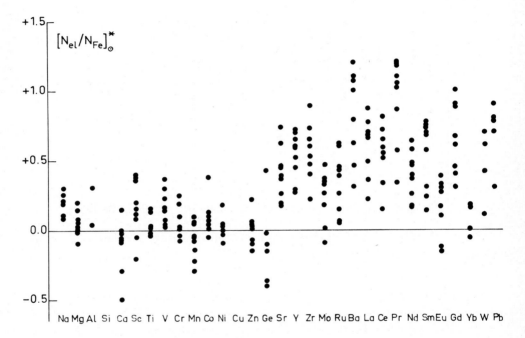

Fig. 9.   Logarithmic abundances $\left[\dfrac{N_{el}}{N_{Fe}}\right]^{*}_{\odot}$ of the elements measured in the Barium stars.

right in judging the blue region for such kinds of stars, with weak lines and enhanced ionized lines, as belonging to much less advanced (sometimes more than by a spectral class) dwarfs.

    There is another much more subtle problem of slightly incorrect classification. Chiefly it concerns the late F and G type stars. In the catalogue many of them have been classified as dwarfs. Many of them belong to Woolley's catalogue of stars nearer than 25 parsecs (Woolley *et al*, 1970). It was easy to put these stars on an $(M_{bol} - \log T_{eff})$ diagram (see Figure 2 in Cayrel de Strobel, 1974). In this diagram filled circles represent dwarfs and open circles subgiants. It is easy to see that many filled circles (dwarfs) seem to be already evolved subgiants.

TABLE I

Change of spectral type for very metal-deficient stars

HD	$V(\text{km}^{-1})$	[Fe/H]$_\odot^*$	Sp.T. from Catalogue		Sp.T. from detailed analysis	
122563	−206	−2.70	GO	IV	K3	II
165195	− 99	−2.70	(G5)		K4	III
221170	−139	−2.70	G2	IV	K4-5	III-II
140283	−107	−2.15	A8	V	F8	VI
128279	−	−2.05	GO	V	G7	IV-III
M92 III-13	−	−2.00	−		K4-5	III-II
19445	−122	−1.75	A4p		F9	VI
25329	−184	−1.64	K1	V	K3	V
2665	+ 56	−1.56	G5	III	KO	II
103095	−150	−1.50	G8	Vp	G9	VI
201626	− 72	−1.45	−		G9	III
219617	−258	−1.40	F8	IV	GO	VI
232078	−392	−1.30	K3	IIp	K4-5	III-II
M13 140	−	−1.20	−		K4-5	III-II
6755	−344	−1.04	F8	V	G8	III
6833	−203	−0.85	G8	III	K2-3	II
26	−372	−0.67	G4	Vp	KO	II
180928	−206	(−0.60)	−		K5	III
41312	−130	−0.60	gK3		K4	II
175329	− 70	−0.58	K1	III-IV	K5	II
5780	−206	−0.43	K4	III	K5	II

5. Comparison between Observed and Theoretical $M_{bol} - \log T_{eff}$ Diagrams

Apart from this discussion of problems of classification, we should like to illustrate the application of well-determined chemical and physical parameters of nearby stars to the construction of HR diagrams in the $(M_{bol} - \log T_{eff})$ plane.

On the $(M_{bol} - \log T_{eff})$ diagram which we have already considered (Cayrel de Strobel, 1974) we can see that many of the observed points towards the bottom fall below the main-sequence. The question was: either the abscissae and the ordinates of these stars are not well-determined, or the slope of the observational main-sequence is different from the theoretical one. On Figure 12 an $(M_{bol} - \log T_{eff})$ diagram has been plotted composed only of stars with normal (solar) metal abundances having both good bolometric magnitudes and good effective temperatures. On the second diagram of Figure 12 the stars of the observational main-sequence fall almost all on a theoretical zero-age-main-sequence calculated by Hejlesen for normal metal/hydrogen abundances (X = 0.70, Z = 0.02) (Hejlesen *et al.*, 1974).

This seems to indicate that the lower observational main-sequence has the same slope as the theoretical main-sequence of Hejlesen.

The broken lines in Figure 12 represent Hejlesen's grid of evolutionary tracks for (X=

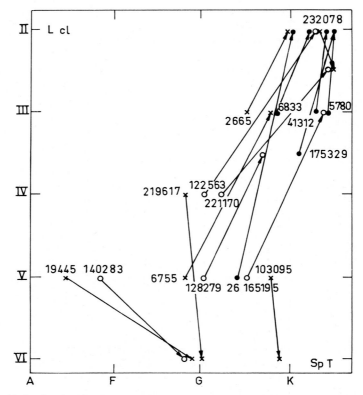

Fig. 10. In this luminosity class vs spectral type diagram arrows connect the two classifications given for each star in columns 4 and 5 of Table I. Open circles, crosses and filled circles respectively stand for stars with the following deficiencies:

$$\circ \quad -2.70 \leqslant \left[\frac{N_{Fe}}{N_H}\right]^{*}_{\odot} \leqslant -2.0$$

$$\times \quad -1.75 \leqslant \left[\frac{N_{Fe}}{N_H}\right]^{*}_{\odot} \leqslant -1.00$$

$$\bullet \quad -0.85 \leqslant \left[\frac{N_{Fe}}{N_H}\right]^{*}_{\odot} \leqslant -0.43$$

0.70 and Z= 0.03) and the solid lines represent a grid of isochrones. The diagram of Figure 12 confirms that of Cayrel de Strobel (1974) concerning the state of evolution of stars classified as dwarfs in our catalogue. On Figure 12 we can see that many of them (filled circles) are already in the subgiant region.

On the (M_{bol} −log T_{eff}) diagram of Figure 13 we have placed nearby metal-deficient stars having good bolometric magnitudes and effective temperatures. Comparing this diagram with a theoretical one calculated by Hejlesen for about the same metal deficiency (X= 0.70 Z = 0.004), we can see that the nearby metal-deficient stars fall already in the evolved subgiant region of the HR diagram with the exception of the cool subdwarf Groombridge 1830. Two well-known subdwarfs lie on this diagram: HD 19445 and HD

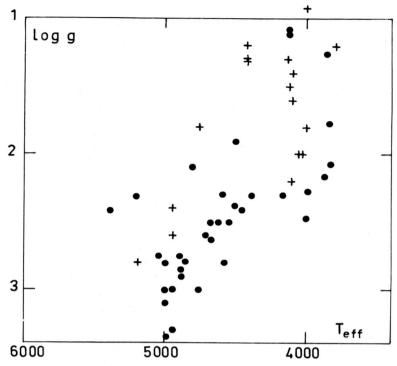

Fig. 11. In this log g vs T_{eff} diagram filled circles stand for population I giants with spectral type between G5 to K5. Crosses stand for metal-deficient stars contained in Table I and falling in the same range of temperature and gravities as the population I giants. Here it is visible that the temperatures of the metal-deficient stars correspond to late G and early K stars, their gravities being on average smaller than those of population I yellow giants.

140283. These stars are more metal-deficient than $Z = 0.004$, nevertheless even on this mild metal-deficient theoretical HR diagram HD 19445 and HD 140283 have already passed the turn-off point as suggested by R. Cayrel (1968).

The nearby subgiants on the HR diagrams of Figures 12 and 13 can be used also to test the age of the stars in the solar neighbourhood.

6. Conclusion

We said in the introduction that a metal/hydrogen abundance catalogue of stars could be useful to astronomers in their work.

We have shown the metal abundance diagrams of individual elements, but we leave the interpretation of these diagrams to astronomers interested in nucleosynthesis or in diffusion problems.

Note, that the problem of the Hyades remains acute since the high dispersion analyses

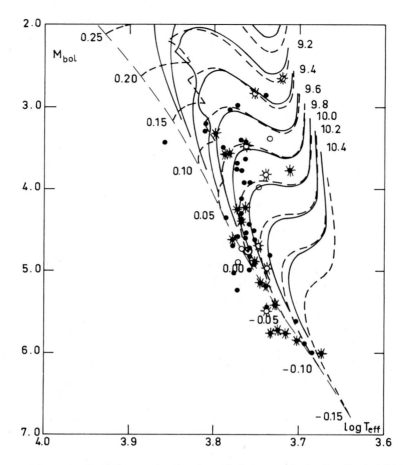

Fig. 12. The (M_{bol} − log T_{eff}) diagram showing the evolutionary tracks (broken lines) and isochrones (continuous lines) computed by Hejlesen for normal metal abundances ($X = 0.70, Z = 0.03$) together with the positions of stars with normal metal abundances, good bolometric magnitude and good effective temperature determinations. The full circles in the diagram are stars classified as dwarfs, in the catalogue, the open circles are stars classified as subgiants. Full circles with asterisks and open circles with asterisks are stars having very well-determined bolometric magnitudes. The open dotted circle shows the position of the Sun. The ages are given in log years; the masses in log M/M⊙.

lead to normal metal abundances in contradiction with the results of Strömgren's, Gyldenkerne's, Geneva and other photometries, which give to the Hyades a metal/hydrogen abundance which is approximately twice the value of the normal 'solar' one.

Let us conclude this introduction to the metal/hydrogen catalogue with the following remarks.

Out of 343 stars contained in this catalogue with spectral types more advanced than F5 only 6 have metal-deficiency factors greater than 100, with respect to the sun. Five were discovered some twenty years ago by Greenstein and co-workers. Since then many

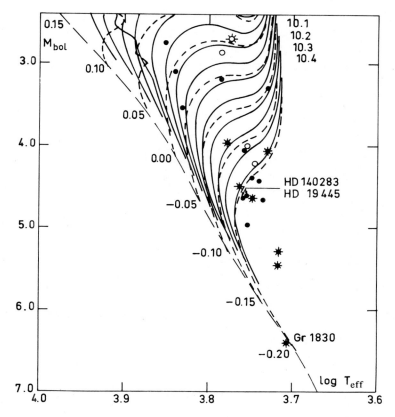

Fig. 13. The (M_{bol} − log T_{eff}) diagram showing the evolutionary tracks (broken lines) and isochrones (continuous lines) computed by Hejlesen for weak metal abundances ($X = 0.70$, $Z = 0.004$) together with the positions of stars with weak metal abundances, good bolometric magnitude and good effective temperature determinations. Full circles, full circles with asterisks, open circles and open circles with asterisks have the same meaning as in Figure 12.

high-velocity stars have been investigated, only one of them has been found as metal-deficient as the Greenstein stars (this star was discovered by the Spites at ESO Observatory two years ago). Many of the high-velocity stars have been found only slightly metal-deficient.

This is an important result and it leads to the conclusion that the halo is probably considerably less metal-deficient than is currently assumed. The photometry of Steinlin (1975) (Becher's system) in Basle contradicts this statement. This discrepancy is likely explainable by the difference in the criteria used to define a 'halo' star. Our definition based on the modulus of the space velocity (U, V, W) includes stars which may belong to the disk population and which should be better described as 'high velocity stars', whereas Steinlin has a very pure sample of halo stars, on the ground of their distance to the galactic plane.

It would be extremely interesting to carry out a detailed analysis of a sample of Steinlin stars of magnitudes $\simeq$ 15 to 16. That does not seem to be out of reach of modern observational techniques. I hope that a second generation 'Catalogue of iron/hydrogen determinations' will contain a large sample of faint stars.

References

Blackwell, D. E.: 1974, *Quart. J. Roy. Astron. Soc.* **15**, 224.
Böhm-Vitense, E.: 1975, in B. Baschek, W. H. Kegel, and G. Traving (eds.) *Problems in Stellar Atmospheres and Envelopes,* Springer-Verlag.
Carbon, D. F. and Gingerich, O.: 1969, in O. Gingerich (ed.), *Theory and Observation of Normal Stellar Atmospheres,* M.I.T. Press, Cambridge.
Cayrel, R.: 1968, *Astrophys. J.* **151**, 997.
Cayrel de Strobel, G.: 1974, Joint Discussion on Kinematics and Ages of the Stars Near the Sun in *Highlights of Astronomy,* Vol. III.
Gustafsson, B.: 1975, *Astron. Astrophys.* **42**, 407.
Hejlesen, P. M., Jorgensen, H. E., Otzen Petersen, J., and Romcke, L.: 1972, in G. Cayrel de Strobel and A. M. Delplace (eds.), *'L'âge des étoiles', IAU Coll.* **17**, Paper XVII.
Kurucz, R. L., Peytremann, E., and Avrett, E. H.: 1973, *Blanketed Model Atmospheres for Early-type Stars,* Smithsonian Institution Press, Washington D. C.
Mäckle, R., Holweger, H., Griffin, R., and Griffin, R.: 1975, *Astron. Astrophys.* **38**, 239.
Mihalas, D.: 1966, *Astrophys. J. Suppl.* **13**, 1.
Peytremann, E.: 1974, *Astron. Astrophys. Suppl.* **18**, 81.
Steinlin, U.: 1975, private communication.
Warner, B.: 1965, *Monthly Notices Roy. Astron. Soc.* **129**, 263.

DISCUSSION

Bidelman: With respect to the spectral-type vs the *I-K* colours, where are the types from? Are they all MK types, or partly HD?

Also, is it not possible that some of the scatter in this diagram is due to the effect of interstellar reddening? Many of the stars in the *Caltech Catalogue* are barely interstellar-reddened. Perhaps, however, your stars are sufficiently near that there is no problem here.

Cayrel: The stars of the diagram have mostly all MK types. I do not think that the scatter is due to interstellar reddening; the stars are all sufficiently near to permit us to exclude the interstellar-reddening as the cause of the scatter. This scatter could be attributed partially to global abundance effects – or to peculiar abundances effects.

Walborn: The difference between *I-K* scatter for stars with spectral types and with quantitative analysis *could* be due to reddening, as pointed out by Dr Bidelman, particularly if the quantitatively analyzed stars are systematically brighter. If discrepancies exist when only reliable, homogeneous types are used, and reddening is corrected for, then one may be finding interesting objects, with infrared excesses or some other real colour-spectrum discrepancy.

Müller: In the table you showed us you had two columns concerning spectral types, one was the spectral types given in catalogues, the other was derived from 'detailed analysis'. (1) How certain are you of the spectral type derived from detailed analysis? (2) 'Detailed analysis' means 'differential curve of growth' analysis? If so, then this is rather dangerous to do as Dr Baschek already pointed out this morning.

Cayrel: (1) I do not derive spectral types from detailed analysis. By spectral type from detailed analysis I mean the spectral type of a Population I star, with normal 'solar' metal content, having the same effective temperature and the same gravity as the analyzed Population II star.

(2) Yes, detailed analysis means here differential curve of growth analysis. The older iron abundance results: $\left[\dfrac{Fe}{H}\right]\dfrac{\text{star}}{\text{standard}}$, come mostly from differential curve of growth analysis relying upon the one-layer approximation, the recent results rely upon appropriate model-atmosphere computation.

C. Jaschek: Since the Jaschek-Jaschek catalogue was mentioned several times I would like to explain what we did. We chose from the existing literature what we considered the 'best' classification, in the case several classifications were available; in the case only one existed, this single one was taken. The 'best' classification was considered the one made either by Morgan or Keenan, or one of his associates. Objective prism classifications were discarded, if possible. The classifications are therefore definitely not homogeneous.

One very important point is that the spectral classification used be homogeneous; those in the catalogue are definitely not homogeneous. Therefore one should be very critical where large discrepancies appear between these classifications and theory.

Cayrel: (1) Considering *only* in the catalogue a sample of homogeneous spectral MK classification of metal-deficient stars, we have even so found inconsistency between spectral classification and atmospheric parameters.

(2) The differences of the values of the atmospheric parameters between different authors for a given very metal-deficient star are always small when compared to the differences between the spectral type deduced from MK classification and the spectral type attributed to the star from its atmospheric parameters.

Baschek: Which are, in detail, the criteria for luminosity used for the column 'catalogue'?

Bidelman: The criteria of luminosity in the G and K stars are mainly atomic lines — Sr II, Ti II, Fe II, and the hydrogen lines. The molecular bands are not used because they are known to be subject to non-understood variations.

Baschek: Is it possible that a large part of the scatter in luminosity is due to deviations of the solar relative abundances when strontium, cyanogen, iron, ... is involved in the criteria?

Cayrel: Yes, this could be very possible and I answered in this way to Dr Bidelman.

Hack: I agree that a detailed analysis is necessary for reliable spectral classification because it is obviously meaningless to try to fit stars in a two-dimensional classification when the chemical composition is appreciably different from the solar one.

Keenan: The effect of the choice of criteria and the variation of abundances in producing the scatter in Mme Cayrel's diagrams can be judged better after some of the later papers, and can be better discussed then.

Garrison: To illustrate with an example: HD 122563 has an extreme difference between types given by spectral classification and Mme Cayrel's type. It puzzles me because the H lines for a K star are very markedly weaker than a G0 star. Therefore the classifier would have to affix a 'peculiar' to the classification of a star with metal lines that weak, because, unless it is He rich, the H weakness and the metals weakness are inconsistent.

Walborn: It appears that a number of features in several of the diagrams could be due to systematic errors in the bolometric corrections, absolute-magnitude scale, and/or effective-temperature calibration for late-type stars.

Metal deficiencies as large as those in your table should be detected in careful spectral classification, for instance as discrepancies between the G-band-to-hydrogen ratio and the metal-to-hydrogen ratios. It is important not to take spectral types uncritically from large compilations without regard for their sources or quality.

Cayrel: Yes, metal deficiencies can be detectable in other careful spectral classifications, but they have not been detected in 'careful' MK classifications just because the blue spectral region of a metal-deficient star looks very much like the blue spectral region of a hotter metal-normal star.

C. Jaschek: The most probable error is the one in parallaxes — one is not sure that the errors of the parallaxes are smaller than 10%; and if this is not true, the absolute magnitudes may be off by large amounts.

Cayrel: The stars shown on diagrams Figure 12 and Figure 13 of my paper are all nearby stars with very good parallaxes.

Spinrad: Since most stars near the sun are M dwarfs and most of the mass of the Universe is in M dwarfs, it is a pity that the Catalogue does not include any. We badly need to study the physics and abundances of M dwarfs – and with the astrophysical capability of the 1970s this should be possible.

Cayrel: There are two causes which even in the 1970s have excluded M dwarfs from a catalogue of iron/hydrogen determinations, based on high dispersion spectra and on model atmospheres analyses: (a) The iron lines are badly blended and the non-existing continuum prevents equivalent widths measurements.
(b) Consistent model atmospheres for M dwarfs, taking into account appropriate molecular blanketing and appropriate convection, do not exist.

Kandel: As a theoretician (Commission 36) I am inclined to accept an observer's spectral classification. Has one attempted to synthesize a Population II spectrum corresponding to a specific T_{eff} and g, and present it to an observer in a form (tracing, or better still photographic plate, synthesized with the appropriate resolution) such that classification can be made? Is it true in fact that a predicted Population II spectrum could be classified with so much 'error'?

Cayrel: Such a synthetic spectrum, as you describe it, has not been realized. Even so, the blue spectral region of a very metal-deficient Population II synthetic star would appear as a blue spectral region of a much less advanced Population I star to a trained eye of a spectral MK classifier.

In the name of the Organizing Committee, *C. Jaschek* explained briefly the reasons for having a discussion on standard stars. Dr. *A. Maeder* suggested that 'it would be important to discuss the set-up of a list of standard stars, to be used for the calibration of the different photometric and spectro-photometric procedures used to derive abundance of groups of elements'.

Because of this suggestion, a Committee was appointed to set up such a list, of which Mrs *G. Cayrel* became the chairman.

A list of standard stars was proposed by Mrs *Cayrel*.

In the discussion which followed, the following points were raised:
(a) no very bright stars should be included in the list because of the difficulty of observing them photometrically
(b) stars whose peculiarity is not firmly established should be avoided.

As a result of the discussion, it was agreed that the following list could be provisionally used:

886	B2IV
10307	G2V
27749	Am
28068	G1V
28344	G2V
61421	F5 IV–V
62509	KOIII
90537	G8 III–IV
103095	G8Vp
113226	G8III
122563	GOIV
149438	BOV
172167	AOV
182572	G8IV
214680	O9V
219134	K3V

It was further recommended that any additional suggestions should be sent to Mrs *Cayrel* before the end of 1975 and that a discussion on this point should be made at the next meeting of IAU Commission 45, if possible in connection with Commission 29.

PART III

DERIVATION OF ABUNDANCES THROUGH
PHOTOMETRIC AND SPECTROSCOPIC METHODS

DERIVATION OF ABUNDANCES THROUGH PHOTOMETRIC AND SPECTROSCOPIC METHODS

R. A. BELL

Astronomy Program, University of Maryland, U.S.A.

Abstract. Several points of detail which affect stellar abundance determinations are discussed. In particular, the importance of including the effects of hyperfine structure and isotopic shifts when considering the lines of some elements is stressed.

The abundance determinations for F dwarfs by Bell and Peytremann, who use theoretical calibrations for intermediate band photometry, and Nissen, who observes very narrow spectral intervals, are intercompared. The agreement between Bell and Nissen, who have 46 stars in common, is quite satisfactory.

Recent work on carbon and nitrogen abundances in cool stars is described. The suggestion of Hearnshaw, that [C/H] = 1.5 [Fe/H] for disc stars with $-0.7 <$ [Fe/H] < 0.4, is compared with recent results by Clegg. Whilst Clegg's results are quite precise, they neither confirm nor deny Hearnshaw's suggestion. Work by Branch and Bell on K giants shows that [C/Fe] = 0, or a constant, for the stars in the sample. A value of about 7 for the C^{12}/C^{13} ratio in the atmosphere of Arcturus has now been confirmed by several authors and Lambert and his collaborators have determined this ratio for several K giants.

The suggestion by Spinrad, Taylor and others that the M67 dwarfs are more metal-rich than the Hyades, i.e. that they are super-metal-rich or SMR, seems to be erroneous. However some SMR stars, such as 31 Aql, certainly exist even though there is still some uncertainty in the precise abundance of strong CN stars such as μ Leo.

Examples of synthetic spectra for metal-deficient giant stars are given and a theoretical colour-colour diagram is compared with observations of globular cluster and Draco stars.

By now, any discussion of the derivation of stellar abundances must cover an enormous amount of ground. In my talk today I will take my lead from the title and primarily discuss the stars whose abundances can be found by both spectroscopic and photometric means. This means that I'll primarily restrict myself to discussing stars of spectral type F and later. In fact, I will be mainly talking about F dwarfs and K giants. Also, I think that all of the results which I will quote, whether my own or those of others, have been obtained by LTE physics. Tremendous progress has been made in studying the spectra of hot stars, with the problem of obtaining self-consistent solutions of the coupled transfer and statistical equilibrium equations having been solved by Auer and Mihalas (1969, 1973). In addition to discussing various abundance and line profile problems, Mihalas (1972) has also published a treatment of non-LTE effects on the continuum and hydrogen line profiles in OB stars and has discussed the *uvby* colours of these models.

My plan, then, is to mention a few points of detail which affect abundance determinations. I'll then discuss the determination of abundances photometrically for F dwarfs. I will make some comparison between different photometric calibrations and with results found spectroscopically. Then I will discuss recent work on K giants, concentrating on CNO abundances and C^{12}/C^{13} work. If time permits, I hope to mention some work on

B. Hauck and P. C. Keenan (eds.), Abundance Effects in Classification, 49–65. All Rights Reserved.

Pop II giants which I have been carrying out in collaboration with Dickens (Herstmon-ceux) and Gustafsson (Uppsala). As you have noted by now, there hasn't been any mention of the effect of stellar abundances on spectral classification. I will show a few slides to illustrate how the spectrum of a star changes with changes in abundance and gravity. Finally, a few requests, comments and complaints.

The standard method of determining abundances by spectroscopic means requires the measurement of the equivalent widths of individual spectral lines. This forces the use of high dispersion spectra, in order to be able to locate the continuum properly and to minimise the problems with line blending. Given the equivalent widths, the actual deter-mination of the abundances can be carried out very quickly. Apart from any problems of line identification, oscillator strengths and so forth, the major criticism of this approach is the amount of telescope time required to obtain the data. We certainly have to use this approach to analyse suitable bright stars in the solar neighbourhood. However, in many interesting cases, such as stars in external galaxies, it is impossible to obtain the relevant data.

And now for the details. In both spectroscopic and photometric work it is necessary to make accurate estimates of stellar temperatures, in order to obtain accurate abundances. The situation here seems to be fairly satisfactory for cool stars, with a large quantity of $H\beta$ photometry for F stars and $R-I$ for G and K stars being available. We obviously ha-ven't heard the last word on how to convert $H\beta$ and $R-I$ to T_{eff} yet, though. We still find it necessary in many cases to include the microturbulent component of the Doppler Broa-dening Velocity (DBV). Recent solar analyses, however, do indicate that the microturbu-lent velocity is very small for the Sun. For example, Foy (1972) finds that it is only 0.5 km s^{-1} at the centre of the disc. The larger values found in the past for the Sun seem to have been caused by errors in the oscillator strengths or by damping effects, both of which affect the shape of the curves of growth. The work of Oinas (1974), with the alarming suggestion that the continuous absorption coefficient is being greatly under-estimated for K stars, has been challenged by Perrin, Cayrel and Cayrel (1975). These authors argue that the difficulty which Oinas found in reconciling the abundances determined from neutral lines with those found from ionized lines arose from his treat-ment of the damping. I'm rather relieved that Oinas seems to be wrong, especially after several authors including Gustafsson, Eriksson, Nordlund and I have spent so much time computing flux constant model atmospheres without inventing new opacities. As far as the damping is concerned, only for a few lines do we have a satisfactory treat-ment and most authors (e.g. Holweger, 1972) are using the Unsöld treatment with an enhancement factor included. This is, typically, of the order of two or three. The use of this treatment in curve-of-growth calculations leads to curves of growth which depend on excitation potential and we can no longer talk about THE solar curve of growth, since there are an infinite number.

In the past, we have generally neglected the effects of hyperfine structure in the calculation of line absorption coefficients, with the exception of the element manganese. We have also generally neglected isotopic shifts. We must be very careful on these two points when determining the abundances of some elements. For example, Hauge (1972) has given a detailed analysis of the 4205 line of Eu II in the solar spectrum. This line is produced by Eu^{151} and Eu^{153} and each of these isotopes produces a line consisting of six hyperfine components. In other words, when analysing the λ 4205 line in stellar spectra, we must consider 12 lines and not just 1. The separation of the individual components is very small, being about 0.03 Å for Eu^{153} but this must be compared with the Doppler width, which is about 0.02 Å for the Sun. This means that the equivalent width of an Eu II line will be computed to be much greater when the hyperfine and isotopic effects are included compared to when they are not. Hartoog et al. (1974) have shown that allowance for these effects reduces the Eu abundance obtained by as much as 0.9 dex compared with the abundance obtained when the effects are neglected. Allen and Cowley (1974) have similarly shown that the abundance of Pr in Ba II stars is reduced when the effect of hyperfine structure is included and the resultant Pr abundance becomes in much better accord with the predictions of the s process. Holweger and Muller (1974) have discussed the isotopic and hyperfine effects when considering the strengths of the solar Ba II lines.

Any comparison of abundances determined spectroscopically with those found photo-metrically must make allowance for the fact that spectroscopic abundance determinations refer to particular elements whereas photometric determinations refer to an average, or overall, abundance. In what follows, I'll use the notation [Fe/H], [Mn/H],... to refer to spectroscopic results and [M/H]: X, Y to photometric ones, i.e. the average metal abundance obtained from the colours X and Y. Then [A/H] can be used to refer to abundances used in model atmosphere calculations.

The calibration of photometric colours or indices in terms of abundances can be carried out by two completely different methods. Suppose we consider, as an example, the determination of abundances for F dwarfs using $U-B$ and $B-V$.

The original method is to plot the abundance index $(U-B)$ versus the temperature index $(B-V)$, measure the deviation of the abundance index for a particular star from the mean line $(\delta(U-B))$ and plot this deviation vs [Fe/H]. The resultant figure supplies the calibration, usually expressed as an equation. The choice of [Fe/H] as the abundance parameter seems to be the wisest one in the circumstances, since Fe supplies so much of the solar line blocking. Moreover, most of the metal abundances seem to vary pretty well in unison and we do not yet have to cope with a situation where a star is deficient in iron yet overabundant in, say, nickel.

The calibration obtained by this method is very useful and it certainly behoves anyone who has found that a star is metal-deficient to check whether it has an ultra-violet excess. Nevertheless, there are a number of problems with this approach. Firstly, there is the question of the influence of the stellar gravity on the colours. Secondly, the ultra-violet excess is not produced solely by iron and there is the possibility that some particular element may be particularly effective at producing ultra-violet excesses. An example of such an element is carbon, which can produce so many molecular lines. Thirdly, we expect that a given abundance change will produce different ultra-violet excesses in stars of different temperatures. The fourth possibility, frequently mentioned a few years ago, that observed ultra-violet excesses can be produced by differences in microturbulent velocity, need no longer be considered. Whilst all of these problems can be checked observationally, it is a big job to do so, especially since we really need homogeneous results.

The alternative, and more recent, calibration method relies on the use of synthetic stellar spectra. These spectra are computed using model atmospheres and large quantities of atomic and molecular line data and are then convolved with photometric sensitivity functions. A significant number of spectra can be computed with a wavelength resolution of 0.1 Å covering the wavelength interval between 3000 Å and 12 000 Å without requiring the expenditure of excessive computer time. This resolution is sufficient for most colour calculations. By carrying out the calculations for different abundances, a theoretical calibration of the various indices can be obtained.

Usually, an essential point in this work is the determination of the zero points and scale factors for the theoretical colours. For example, if we denote the initially calculated $U-B$ colour as $(U-B)_c$ then the colour which we wish to compare with the observations, $(U-B)_*$, may be given by $(U-B)_* = a (U-B)_c + b$. Hopefully, the sensitivity functions used in the convolution closely resemble the actual ones, so that $a = 1$, but there is still the problem of determining b. This problem can be solved by identifying a particular star with a particular model. We then say that the colours of the model are the colours of the star. This calibration star is then serving the same purpose as the reference star in differential curve-of-growth analyses. Carrying out the calculations for more than one colour system has advantages here, in that it can point out if inappropriate models are being used. As an example of this, I refer to recent calculations Gustafsson and I made. The colour differences between φ^2 Ori and the Sun match the computed colour differences much better if we say that φ^2 Ori has $T_{eff} = 4600$ K instead of, say, 4500 K. At the present time, I do not think we can expect that one reference star will be sufficient for all purposes. For example, if we are working on F and G dwarfs then we can use the Sun as the reference star whereas for cooler, lower gravity and more metal-deficient stars we might well use HR 1907 (φ^2 Ori). One point which I think could well be pursued in more detail at this meeting concerns the establishment of a network of such calibration stars which must be observed by all individuals who are establishing a new system. Even if such a system is intended to refer to K giants, it might be valuable to have observations of B stars to aid in subsequent interpretation. Any network of calibration stars should include a substitute Sun, since the genuine article is so difficult to observe accurately.

Returning to the calibration which is obtained from synthetic spectra, the most serious criticism which can be made is that there is no guarantee that the results will be correct, at least at the present time. As the relevant synthetic spectra become in better and better agreement with the best observed spectra, we can have more and more confidence in the calculations. The results depend heavily on the atomic and molecular line data and, to a lesser amount, on the model atmospheres. For these reasons, the colour predictions must be carefully checked, i.e. if the models do not predict a variation of $\delta(U-B)$ with [M/H] which resembles the observed variation, then one must be sceptical about the ultra-violet fluxes and about other ultra-violet colours. However, one must also be very cautious about the possible influence of particular elements on the colours. For example, one should bear in mind the possible problem with carbon particularly in view of Hearnshaw's (1974a,b, 1975) work.

At this point it is tempting to compare the results which have been obtained by different workers for this problem. Peytremann (1975) has recently published a discussion of the Geneva photometry based upon his flux constant models and synthetic spectra. I will use some of the results which he gives for a sample of G2 V stars. Peytremann's approach is a little different from that outlined above, in that his synthetic spectra are computed with many fewer wavelength points. He also neglects molecular lines. As a foil for this work, I will use my own results for these stars. My calibration is based upon the use of scaled solar model atmospheres and high resolution synthetic spectra (Bell, 1970, 1971) and I have used $uvby$ photometry as the observational data. In addition to estimating [M/H], both Peytremann and I have estimated T_{eff} and log g. I have used Hβ and $R–I$ to get T_{eff}, c_1 and $b–y$ to get log g and m_1 and $b–y$ to get [M/H]. Peytremann has tried to find the values of T_{eff}, log g and [M/H] which give the best agreement between all the observed colours and all the computed ones. The results of the comparison for the stars in common are given in Table I. I'll leave it to the reader to say if he considers the comparison to be encouraging or not. However, to guide the reader, I should point out that Peytremann states that the Geneva system is not particularly good for finding log g for solar type stars. As far as my results go, Hβ data loses a lot of its precision as a T_{eff} indicator at about 5800 K and for two of the stars in the Table, 39 Ser and 85 Peg, my calibration of $B–V$ indicates a rather higher value of T_{eff}. There is a rather curious point concerning the results of Table I, [M/H]: m_1, $b–y$ – [M/H]: Geneva

TABLE I

Comparison of T_{eff}, [M/H] and g found from intermediate band photometry

Star	T_{eff}		[M/H]		log g	
HR/Name	B	P	B	P	B	P
483/	5800	5840	+.26	+0.2	4.0	4.3
3951/20 LMi	5730	5770	+.57	+0.6	3.5	4.2
5911/39 Ser	5300	5580	−0.30	−0.7	4.7	4.2
5968/ρ CrB	5540	5680	−0.10	−0.4	4.0	4.1
6458/72 Her	5540	5500	−0.22	−0.6	3.9	4.0
7503/16 Cyg	5650	5630	+0.39	+0.0	3.6	4.0
7569/	5600	5690	−0.02	+0.1	3.4	4.0
9088/85 Peg	5040	5180	−0.78	−1.2	4.9	4.1

B = Derived by Bell (1971) from u v b y photometry
P = Derived by Peytremann (1975) from Geneva photometry

is either small or about 0.4. Whilst this Symposium is devoted to abundance problems, I would also like to quote predictions of gravities as these can be compared with 'observed' gravities deduced from assumed masses and radii, the latter coming from M_v and T_{eff}. If the predicted and observed gravities agree, we can have that much more confidence in the abundances.

Some very interesting studies on the abundances of F dwarfs have been carried out recently by Nissen (1970) and Gustafsson and Nissen (1972). These authors have measured the radiation from stars in very narrow bands, Nissen using 3.5 Å, carefully chosen to contain suitable lines. One of the bands contains weak lines, sensitive to [Fe/H], whilst another, containing very weak lines, serves as a reference. For giants, another band will contain lines on the flat part of the curve of growth, sensitive to the DBV. The measurements are made photoelectrically. It is necessary to allow for the radial velocity of the star at the time of observation when setting up the equipment. I was a little startled to learn that the precision of the latest spectrometer is such that the Danish group could consider observing $C^{13}H$ lines in the spectra of K giants. Synthetic spectra are used to calibrate the relative intensities of the bands in terms of [Fe/H] and, if need be, DBV. The results obtained by this method are probably very reliable, at least for stars which have weaker spectral lines than does the calibrating star. In this context, the calibrating star is the one for which very high dispersion spectra are available and which can be used to check the adequacy of the line list used for the synthetic spectra calculations. If we study stars which are more metal rich or cooler than the calibrating star, then lines may appear in the stellar spectra which are not in the line list. This will yield an erroneous calibration. This is especially true for the cool giants, for which this method has also been used. Gustafsson *et al.* (1974) (hereafter GKA) had to go to a great deal of trouble to ensure that they had included all possible lines in their pass bands. Inclusion of the red system of CN is particularly important owing to the large number of possible bands. For example, Griffin's (1970) Sc index measurements could not be used for Sc abundance determinations owing to the contamination by red CN. Returning to the Nissen work, it seems to me that one of the real advantages of this method is that there is no problem of continuum location, which I think bedevils photographic spectroscopic work. Even if one has spectra of sufficient scale, there still remains the problem of grain noise. The problem of continuum location is especially critical for measuring the equivalent widths of weak lines, which are the ones we really want to use for abundance studies. The photoelectric data is much more readily reproducible.

A comparison of the abundances found by Nissen (1970) and by Bell (1971) (from synthetic spectroscopy and *uvby* photometry) shows quite good agreement. For 46 non-cluster stars in common we find [Fe/H]: N − [M/H]: m_1, $b−y$ = +0.09 and the rms deviation is 0.17.

I think that this very narrow band work probably provides the most accurate way of determining [Fe/H] for many kinds of stars and my only reservations concern the possible adequacy of the line list and the relation of the abundances to the Sun. Nissen's abundances depend on the $b−y$ which he used for the Sun and I think his value of 0.424 is about

0.03 mag. too red. This will not affect the scale of his results, only the zero-point. I will return to the GKA results later.

Recent interest in CNO abundances has been, I think, stimulated by work on globular cluster stars such as Zinn's (1973) demonstration that considerable differences exist in the strengths of the G bands in subgiants and asymptotic giant branch stars in M 92. Butler, Carbon and Kraft (1975) have argued that similar differences also exist in the strengths of the NH bands and that the strength of NH is anti-correlated with the strength of the G band. Osborn (1971a) and others have observed a number of globular clusters using the DDO system and Osborn (1971b) pointed out that the anomalous $C(41-42)$ colours of two stars, one in M5 and one in M10, might be caused by differences in CNO from star to star in the cluster. Dickens and Bell (1975) have shown that there is a considerable range in [N/H] in ωCen giants and Bell and Dickens (1974) analyzed the ω Cen CH stars. I will show a slide from this latter paper as it indicates a spectral classification point. In the upper panel of Figure 6 of Bell and Dickens (1974) we see the spectra of two ω Cen Giants, RGO 55 (a CH star) and RGO 84. Note the strength of the Ca I line in the normal giant and its relative weakness in the CH star. In the lower panel we see the synthetic spectra computed for these two objects and it is clear that the enhancement of CH in RGO 55 greatly reduces the contrast in this spectral region and the 4226 line appears much less strong.

Work on CNO abundances in field F, G and K stars has been carried out recently by, amongst others, Sneden (1973, 1974), Hearnshaw (1974a, b) and Clegg (1975). Sneden and Clegg used spectral synthesis to analyse NH, CH and CN whereas Hearnshaw measured equivalent widths of CH lines and analyzed them differentially relative to the Sun, following the treatment in Hearnshaw (1973). It is necessary to be particularly careful when measuring equivalent widths in the neighbourhood of the G band. For the impressively large sample of 50 dwarfs and subgiants between F6 and K0, with $-0.7 <$ [Fe/H] $< +0.4$, Hearnshaw finds [C/H] ~ 1.5 [Fe/H], an extremely interesting result. However in a later paper (Hearnshaw 1975), the result is diluted by the suggestion that it may not be obeyed by high velocity stars. Sneden has considered 11 metal-poor stars and concludes that in metal-poor dwarfs [C/Fe]=0 and, with one exception, [N/Fe]= 0 whereas in giants [C/N] depends on gravity, being -1 if $\log g < 2.4$ and > 0 for higher g. However, Sneden's error bars are typically $\pm$ 0.6: but at least this indicates that departures from the 'null hypothesis' (i.e. [C/H] = [Fe/H]) for C and N are not large. In a sample of 11 dwarfs, Clegg finds deficiencies in a few of the metal-poor stars, indicating that there may be something in Hearnshaw's result. The star α Tri appears to be quite C deficient, a result also found by Pagel. Clegg also finds some N deficiencies in the metal-poor stars: and whilst these are probably real for 4 high-velocity stars μ Cas, 171 Pup, HD184499,85 Peg) they are not as large as the suggested relation [N/Fe] = [Fe/H] would predict.

In an attempt to cast more light on this problem of the carbon abundances, Branch and I have tried to determine carbon abundances for G and K giants. If Hearnshaw is correct, we would expect his result for the dwarfs to also hold for giants. The observational data we have used for this purpose is that of Alexander and Branch (1974). This data consists of photometric measurements through three filters, one centred on the (1,0) sequence of the C_2 Swan bands and a comparison band on either side. The system is shown in Figure 1 of Branch and Bell (1975) where, for comparison purposes, we show the ratio of the fluxes observed by Willstrop (1965) for κ Oph and α Boo. It is clear from this that there is greater absorption centred at 4650 Å in the spectrum of κ Oph and we believe that this is C_2. The system possesses one major advantage. The C_2 bands are composed of many lines each of which is spectroscopically weak and so sensitive to changes in the abundance of C_2. The overall feature thus has the advantage of being strong enough to be readily measureable and yet still being very sensitive to abundance. The complication of course is that there are many other lines contaminating the pass bands. Branch and I carried out the relevant synthetic spectrum calculations (we are very grateful to Dr B. Gustafsson for supplying the scaled solar model atmospheres and to the University of Copenhagen for the computer time) and convolved our spectra with the filter transmission profiles. We then took data on the abundances, effective temperatures and gravities of G and K giants from the literature and computed what the C_2 indices of these stars would be expected to be under the supposition that $[C/H] = [O/H] = [Fe/H]$. The results are shown in Figure 4 of Branch and Bell (1975). The abcissa in each of the three plots is the observed C_2 index of a star and the ordinate is the predicted index. The data for the left-hand panel comes from Williams (1971, 1972), the centre panel uses GKA [Fe/H] and the right-hand panel uses [Fe/H] from Hansen and Kjaergaard (1971) and T_{eff} and g from Williams. We see a correlation in all three panels but with less scatter in the centre and right-hand ones. We ascribe the smaller scatter to more accurate values of [Fe/H] which allows us to compute more accurate C indices. If we change the carbon abundance by relatively small amounts e.g. say to $[C/Fe]=+0.15$ then the predicted C_2 index changes by about 0.07, a very large amount in these diagrams. We conclude that $[C/Fe]=0$ for the K giants in this sample.

When carrying out these calculations we used relatively low values for the solar (i.e. reference) C and O abundances. Repeating the calculations with higher values causes the predicted C_2 indices to become a little weaker owing to greater depletion of C by CO formation with the C_2 electronic oscillator strength being altered to keep the solar C_2 line strengths constant. This effect may have produced the zero-point shift.

In this context, I would like to mention some rather curious points concerning the star ν Indi. Przybylski (1962) obtained a spectrum of this object at Mount Stromlo and pointed out that the CN bands were very weak. Harmer and Pagel analyzed the star and found that the CH lines gave $[C/H] =-1.2$ and then the CN bands implied $[N/H] =-2.0$. At Pagel's suggestion, I used my synthetic spectrum programme to compute synthetic spectra for the star and confirmed the C abundance. However, I found an even lower nitrogen abundance, $[N/H] =-3.0$. Five years, and many synthetic spectra, later, I'm rather sceptical about this result. Firstly, it seems that ν Indi is rather hotter than was thought earlier, owing to errors in the photometry. Secondly, it also seems very likely that it has a much higher gravity than earlier thought. The point here is that the *uvby* photometry of Stokes (1973) shows that ν Indi has a very low value of c_1 for its $b-y$ and, in fact, in Bond's (1971) c_1, $b-y$ diagram ν Indi lies in the same location as the stars BD+66° 268 and BD−0° 4470 which Bond classifies as subdwarfs. The parallax values are 0.046 (Yale) and 0.011 (Cape) and the mean value indicates it is a subgiant. If the star is a subdwarf, as the photometry inplies, then it must have a parallax even greater than the Yale value. We're still a bit uncertain about how good theoretical ultra-violet excesses are, but these calculations certainly support the suggestion of the higher gravity from the $uvby$ data. The higher T and g mean that the CN and NH bands can be fitted with a much higher N abundance than found earlier and I think we can no longer argue that ν Indi definitely has [N/Fe] <0.0. I request that the parallax be re-measured.

Lambert, Dearborn and Sneden (1974) state that their stimulus for the recent work on red giants comes from (i) new observing techniques such as Fourier transform infra-red spectroscopy, (ii) the realization that mixing does occur in stars and can produce changes in surface composition and (iii) developments in model atmosphere calculations. Lambert and his colleague have found C^{12}/C^{13} ratios for numerous stars, using the red CN system, CO bands and the G band (cf. Lambert and Dearborn, 1972; Day et al., 1973; Tomkin and Lambert 1974 and Lambert and Tomkin 1974). This work, and confirming work by other authors (Upson 1973, Krupp 1973 and Griffin 1975) has shown that α Boo has a C^{12}/C^{13} ratio of about 7, compared to the terrestrial/solar ratio of 89. One of the advantages of synthetic spectra calculations is that the observational data can be shown directly and in Figure 1 we see the p_{1dc} (11) line of $C^{13}H$ in the spectrum of α Boo, the data being taken from Krupp (1973). In fact, using Krupp's (1974) precise wavelengths, lines of $C^{13}H$ can be found in the solar spectrum using the Revised Rowland Tables (Moore, Minnaert and Houtgast (1966). However, the solar C^{12}/C^{13} ratio can be more accurately obtained from CO lines (Hall, Noyes and Ayres 1972). In Figure 2 of Lambert and Tomkin we see the observations of ϵ Peg obtained using the Tull coude scanner at MacDonald and it is clear that, from this superb observational material, it is possible to distinguish between C^{12}/C^{13} ratios of 4,5 and 6. In fact, Lambert and Tomkin give 5.1 ± 0.5 as the C^{12}/C^{13} ratio for ϵ Peg. With this high precision, it is necessary to include relatively subtle effects such as explicitly considering the satellite lines and, in fact, we are almost near the level of precision where it is necessary to allow for vibration-rotation interaction.

A number of the results obtained by Lambert's group are shown in Figure 1 of Lambert and Tomkin which gives evolutionary tracks from Paczynski (1970) and the C^{12}/C^{13} ratios for various stars. We see that the SMR star μ Leo doesn't have an especially low ratio. Subgiants have higher C^{12}/C^{13} ratios.

Wing (1974) has described a photometric system for determining C^{12}/C^{13} in K giants, using ~20 Å band passes. We look forward to seeing his results.

What about super-metal-rich (SMR) stars? Even though Spinrad, who introduced this term (Spinrad and Taylor, 1969, hereafter ST), is here I would like to make a few comments since this is one case where the photometric indices appeared to disagree with the spectroscopic ones. To remind you of the situation, ST followed earlier work at

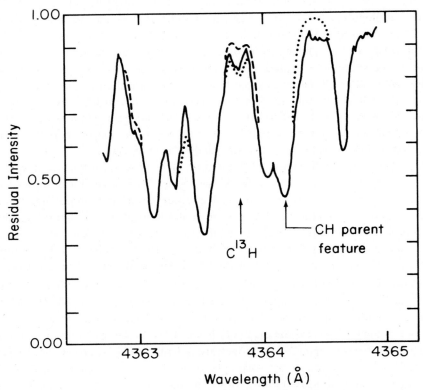

Fig. 1. The observed spectrum of Arcturus (Griffin 1968) is represented by the solid line. The dashed and dotted lines represent synthetic spectra for $C^{12}/^{13}$ = 19 and 9, respectively. The P_{1dc} (11) line of C^{13} H and its parent C^{12} H line are identified.

Cambridge and used band passes of 15 or 30 Å to observe K giants. Some band passes are centred on strong lines or molecular features, others serve as references. On the basis of their observations, ST concluded that 'Evolved K stars with metal abundances greater than those of the Hyades exist in substantial numbers'. This is a controversial result since it is widely felt that, whilst some stars with abundances greater than those of the Hyades do exist, they do not exist in substantial numbers. One of the SMR stars is μ Leo, previously known to be a strong CN star (e.g. Griffin and Redman 1960). ST found [Ca/H] = [Mg/H] = [Na/H] =0.6 for this object. Since ST also claimed to find a N-Na-Fe correlation, μ Leo would also be Fe rich. Pursuing the topic vigorously, ST argued that the M67 dwarfs were SMR (Spinrad *et al.*, 1970) both from D line indices and from low dispersion spectra. Subsequent discussion by Abt and Morgan (1973) and Barry and Cromwell (1974) has indicated that the M67 spectra were not as sharply focussed as usual and better spectra do not reveal the abundance anomaly. No explanation has been subsequently offered, to my knowledge, as to why the D line measures give lines which are stronger than those in corresponding Hyades stars.

More recent spectroscopic and photometric work has not confirmed the ST abundances for μ Leo. Strom, Strom and Carbon (1971) found [Fe/H]=+0.4 (their value is relative to the Hyades which I take to have [Fe/H]=+0.3). Blanc-Vaziaga, Cayrel and Cayrel (1973, hereafter BVCC) obtained [Fe/H]=–0.1 and Pagel (unpublished) states that 'μ Leo and α Ser have very closely similar compositions and their abundance is not greater than that of the Hyades'. Strom *et al.* used 8 Å mm^{-1} spectra, BVCC used 6.8 and 12 Å mm^{-1} whilst Pagel used 2.9 Å mm^{-1}. GKA found [Fe/H]= 0.58 or 0.39, depending on DBV, again adopting 0.3 for the Hyades. I give greatest weight to Pagel and GKA in this because of dispersion and because of method, respectively. If we take the lower GKA value we have consistency between the two methods. One problem affecting the GKA work, which will affect much photometric work, is this difficulty of the extraneous lines. In other words, what is the effect of variations in the abundance of other elements relative to Fe in the final value of [Fe/H]? In their discussion, GKA conclude that the abundance of Mg is particularly important for them, owing to the presence of MgH lines in one of their pass bands as well as the importance of Mg as an electron donor. No spectroscopist seems to find μ Leo to be Mg rich. Lines of CN are equally distributed in both feature and reference bands and thereby the importance of CN variations from star to star is minimised in their case. On the basis of these results we conclude that ST overestimated their Fe abundances. BVCC state that μ Leo may be overabundant in Ca, Na, Mn, Cu and underabundant in Ba.

Gustafsson, Eriksson, Nordlund and myself have computed a large number of model atmospheres for metal-deficient giant stars. The later set of these models includes the opacity effects of CO and the red system of CN as well as atomic lines, using opacity-distribution functions. The earlier set of models were computed using only atomic line ODFs. We have computed synthetic spectra for both sets of models. The synthetic spectrum programme differs a little in the two cases, with the damping wings of strong lines being followed to only $\pm$ 75$\Delta\lambda_D$ from line centre in the earlier case. The later programme allows the damping wings to have any width. The synthetic spectra were also computed using different CNO abundances but the f_{oo} for the molecular bands were adjusted so that the solar molecular bands at the centre of the disc were fitted. Examples of curves of growth and synthetic spectra, as well as theoretical colour-colour diagrams are shown elsewhere (Bell and Gustafsson 1975), using the earlier models and their colours. The scatter in the observed FeI and FeII curves of growth for φ^2 Ori is very small and an excellent fit can be obtained between observation and calculation. However the scatter in the FeI curve of growth, from an 18Åmm^{-1} spectrum, for the M92 star III-13 is very great and the FeII curve of growth is not worth plotting.

In Figure 2, I show the C(42–45), C(45–48) diagram of the DDO system, computed in collaboration with Gustafsson, for the later set of models and using the latest SSG programme and CNO abundances. The lines shown are the ones on which we expect stars in globular clusters with abundances of [M/H] = –3, –2, –1 and –0.5 to lie. The T_{eff} and g values required to construct the lines come from the evolutionary calculations of Rood (1972), being given in Bell and Gustafsson (1975). The observations of cluster stars on

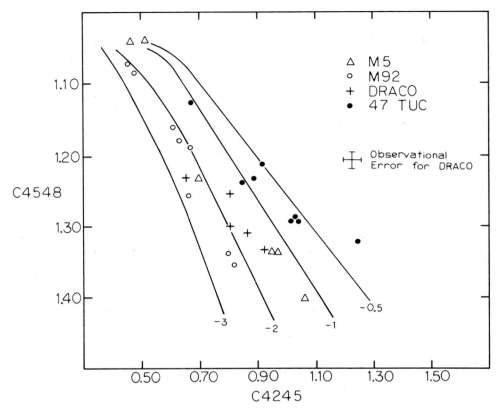

Fig. 2. Using gravities and temperatures found from evolutionary tracks (Rood 1972), theoretical colours on the DDO system are computed for stars in the giant branches of globular clusters with A/H = −3, −2, −1, and −0.5. These are compared with observed C4548, C4245 colours for M5, M92 (Osborn 1973), 47 Tuc, (McClure and Osborn, 1974) and the Draco dwarf spheroidal galaxy (Hartwick and McClure, 1974).

the DDO system (by Osborn, 1971a; McClure and Osborn, 1974; and Hartwick and McClure, 1974) are also plotted in Figure 2. We see that the observations of stars in M92 follow the iso-abundance line for [A/H] = −2 fairly well and, in fact, the average M92 abundance is [M/H] = −2.2. At the present time, with our current equipment, it seems best to try to obtain cluster metal abundances photometrically rather than spectro-scopically, in view of the scatter in the curves of growth. This situation may well change in the future as new equipment becomes available.

I will end this talk with a series of requests.

Firstly, I've found it to be enormously helpful to have high dispersion spectra of stars available for study. Many others, for example photometrists who are deciding which wavelength regions to observe, also need this kind of data. But I would request that it be made available in machine readable form, as well as, or even instead of, a printed atlas. In this connection I've found the Delbouille et al. (1973) solar atlas to be very helpful when computing solar colours. I can convolve the observed solar spectrum with filter sensitivity functions in the same way that I can convolve synthetic spectra. We can also use computer programmes for identifying lines, we can plot the spectra on any desired scale and we can convolve them with any desired instrumental profile if we have data on magnetic tape. If we are to get a really good match between a computed spectrum for a star and the observed spectrum of that star then I think we will just have to do it using a programme which will automatically optimise the agreement between the two.

Secondly, I'd like spectrograph builders and/or users to measure the instrumental profile of their equipment. This is particularly true for the less common instruments, such as the Image Tube Scanners at Lick and the AAT. Much of the data on faint stars is going to be analyzed using synthetic spectra and anyone doing the calculations will need the instrumental data.

Thirdly, let the laboratory spectroscopists continue their good work on both atoms and molecules, with the reminder that we need damping constants as oscillator strengths. Also let Peytremann and Kurucz continue computing their gf values.

Finally, my thanks to all the photometrists and spectroscopists who have obtained all this exciting data on both the mundane and the unusual stars.

Acknowledgments

I am very grateful to Mr John Ohlmacher for all his help over the past several years with the Maryland hardware and software. My thanks also go to Drs B. Strömgren, D. R. Branch, R. E. S. Clegg, R. J. Dickens, B. Gustafsson, P. Kjaergaard, A. Nordlund and C. Perry for so many interesting conversations and ideas.

References

Abt, H. and Morgan, W. W.: 1973, *Astron. J.* 78, 386.
Alexander, J. B. and Branch, D.: 1974, *Monthly Notices Roy. Astron. Soc.* 167, 539.
Allen, M. S. and Cowley, C. R.: 1974, *Astrophys. J.* 190, 601.
Auer, L. H. and Mihalas, D.: 1969, *Astrophys. J.* 158, 641.
Auer, L. H. and Mihalas, D.: 1973, *Astrophys. J. Suppl.* 25, 433.
Barry, D. C. and Cromwell, R. H.: 1974, *Astrophys. J.* 187, 107.
Bell, R. A.: 1970, *Monthly Notices Roy. Astron. Soc.* 154, 343.
Bell, R. A.: 1971, *Monthly Notices Roy. Astron. Soc.* 155, 65.
Bell, R. A. and Dickens, R. J.: 1974, *Monthly Notices Roy. Astron. Soc.* 166, 89.
Bell, R. A. and Gustafsson, B.: 1975, *Dudley Observatory Report* 9, 319.
Bond, H. E. 1971: *Astrophys. J. Suppl.* 22, 117.
Blanc-Vaziaga, M. J., Cayrel, G., and Cayrel, R.: 1973, *Astrophys. J.* 180, 871.
Branch, D. R. and Bell, R. A.: 1975, *Monthly Notices Roy. Astron. Soc.* 173, 299.
Butler, D., Carbon, D., and Kraft, R. P.: 1975, *Bull. Am. Astron. Soc.* 7, 239.
Clegg, R. E. S.: 1975, Ph. D. Thesis, University of Maryland.
Day, R. W., Lambert, D. L., and Sneden, C.: 1973, *Astrophys. J.* 185, 213.
Delbouille, L., Neven, L., and Roland G.: 1973, *Atlas Photométrique du Spectre Solaire de λ 3000 à λ 10000*, Université de Liège.

Dickens, R. J. and Bell, R. A.: 1975, *Bull. Am. Astron. Soc.* **7**, 499.

Foy, R.: 1972, *Astron. Astrophys.* **18**, 26.

Griffin, R. F. and Redman, R. O.: 1960, *Monthly Notices Roy. Astron. Soc.* **120**, 287.

Griffin, R. F.: 1968, *A Photometric Atlas of the Spectrum of Arcturus*, Cambridge Phil. Soc.

Griffin, R. F.: 1970, *Monthly Notices Roy. Astron. Soc.* **147**, 303.

Griffin, R. F.: 1974, *Monthly Notices Roy. Astron. Soc.* **167**, 645.

Gustafsson, B. and Nissen, P. E.: 1972, *Astron. Astrophys.* **19**, 261.

Gustafsson, B., Kjaergaard, P., and Andersen. J.: 1974, *Astron. Astrophys.* **34**, 99.

Hall, D. N. R., Noyes, R. W., and Ayres, T. R.: 1972, *Astrophys. J.* **171**, 615.

Hansen, L. and Kjaergaard, F.: 1971, *Astron. Astrophys.* **15**, 123.

Hartoog, M. R., Cowley, C. R., and Adelman, S. J.: 1974, *Astrophys. J.* **187**, 551.

Hartwick, F. D. A. and McClure, R. D.: 1974, *Astrophys. J.* **193**, 321.

Hauge, O.: 1972, *Solar Phys.* **27**, 286.

Hearnshaw, J. B.: 1973, *Astron. Astrophys.* **28**, 279.

Hearnshaw, J. B.: 1974a, *Astron. Astrophys.* **34**, 363.

Hearnshaw, J. B.: 1974b, *Astron. Astrophys.* **36**, 191.

Hearnshaw, J. B.: 1975, *Astron. Astrophys.* **38**, 271.

Holweger, H.: 1972, *Solar Phys.* **25**, 14.

Holweger, J. and Muller, E.: 1974, *Solar Phys.* **39**, 19.

Krupp, B. M.: 1973, Ph. D. Thesis, University of Maryland.

Krupp, B. M.: 1974, *Astrophys. J.* **189**, 389.

Lambert, D. L. and Dearborn, D. S.: 1972, *Mem. Soc. Roy. Sci. Liège.* 6th Serie, **3**, 147.

Lambert, D. L. and Tomkin, J.: 1974, *Astrophys. J.* **194**, 89.

Lambert, D. L., Dearborn, D. S., and Sneden, C.: 1974, *Astrophys. J.* **193**, 621.

McClure, R. D. and Osborn, W.: 1974, *Astrophys. J.* **189**, 405.

Mihalas, D.: 1972, *Astrophys. J.* **176**, 139.

Moore, C. C., Minnaert, M. H. and Houtgast, J.: 1966, *Natl. Bur. Std. U.S. Monograph* **61**.

Nissen, P. E.: 1970, *Astron. Astrophys.* **6**, 138.

Oinas, V.: 1974, *Astrophys. J. Suppl.* **27**, 391.

Osborn, W.: 1971a, Ph. D. Thesis, Yale University.

Osborn, W.: 1971b, *Observatory*, **91**, 223.

Osborn, W.: 1973, *Astrophys. J.* **186**, 725.

Paczynski, B. J.: 1970, *Acta Astron.* **20**, 47.

Perrin, M. N., Cayrel, R., and Cayrel, G.: 1975, *Astron. Astrophys.* **39**, 97.

Peytremann, E.: 1975, *Astron. Astrophys.* **38**, 417.

Przybylski, A.: 1962, *Publ. Astron. Soc. Pacific* **74**, 230.

Rood, R. T.: 1972, *Astrophys. J.* **177**, 681.

Sneden, C.: 1973, *Astrophys. J.* **184**, 839.

Sneden, C.: 1974, *Astrophys. J.* **189**, 493.

Spinrad, H. and Taylor, B. J.: 1969, *Astrophys. J.* **157**, 1279.

Spinrad, H., Greenstein, J. L., Taylor, B. J., and King, I. R.: 1970, *Astrophys. J.* **162**, 891.

Stokes, N. R.: 1973 (private communication).

Strom, S., Strom, K., and Carbon, D.: 1971, *Astron. Astrophys.* **12**, 177.

Tomkin, J. and Lambert, D. L.: 1974, *Astrophys. J.* **193**, 631.

Upson, W. L.: 1973, Ph. D. Thesis, University of Maryland.

Williams, P. M.: 1971, *Monthly Notice Roy. Astron. Soc.* **153**, 171.

Williams, P. M.: 1972, *Monthly Notice Roy. Astron. Soc.* **158**, 361.

Willstrop, R. V.: 1965, *Mem. Roy. Astron. Soc.* **69**, 83.

Wing, R. F.: 1973 (private communication).

Zinn, R.: 1973, *Astrophys. J.* **182**, 183.

DISCUSSION

Hack: You have shown synthetic spectra at constant T and g and varying chemical composition, and at constant g and composition and varying T. Did you compute also the effect of small changes of microturbulence (by 2 or 3 km s^{-1})?

Did it make changes which can be distinguished from changes in T or chemical composition?

Bell: We have done such calculations but I have not plotted them. I have not analysed them carefully enough to answer your second question.

Nissen: In your calibration of m_1 in terms of metal abundance you assume that the atmospheric velocity parameter — the so-called microturbulence — does not change as a function of metal abundance. The fact that the metal abundances you derive agree very well with those that I have determined from a narrow-band index seems to show that this assumption is correct, because the lines in the v-band of the m_1 index are rather strong and much more sensitive to microturbulence than the weak lines of the narrow-band index.

Bell: Yes, at least using my 1970 calibration.

Foy: I have measured the microturbulent velocity ξ in the solar photosphere by two ways.

(1) From fitting theoretical curves of growth to the observed one for low excitation potential lines of neutral iron, I have found $\xi \simeq 1.0$ km s^{-1} (Foy, 1972).

(2) I have fitted theoretical line profiles to the observed ones for rather weak lines. I have used the solar spectrum Atlas from Delbouille *et al.* (1974) and, as in (1), the solar photosphere model by Peytremann (1974). Results depend upon the lines considered, because this procedure is more sensitive than the first one to the oscillator strengths and to the iron abundance adopted in the computations. Microturbulent velocities which I have found lie between 0.5 to 0.8 or 1.0 km s^{-1} (Foy, 1975); ξ cannot be larger than 1.0 km s^{-1} to fit the data. The observed profiles cannot be interpreted without involving microturbulent velocity: ξ cannot be decreased under 0.5 km/s.

McCarthy: I was interested in the observational data concerning the ^{13}CH feature in Arcturus as shown in your slide. Who identified this feature and with what spectroscopic equipment?

Bell: Lambert and Dearborn initially. Their work was confirmed by Krupp, who re-analyzed the spectrum of C^{13} H to obtain better laboratory wavelengths.

Maeder: Would you please make some comments on the evidence concerning systematic differences of observed surface abundances between the bottom and the top of red giants in clusters. Such evidence would reveal what in the peculiar abundances may be considered as a result of interior evolution and what may be considered as a reflection of peculiar initial abundances, which is a problem related to galactic structure.

Bell: I can only refer you to my earlier comments and the published literature for clusters other than ω Cen. In ω Cen, Dickens and I have spectra of one star with a very strong Ba II λ 4554 line, of other stars with very strong blue CN and of the CH stars. The work on CH stars has been published.

Williams: Your use of a constant Doppler broadening velocity for all stars implies lower microturbulent velocities in the cooler stars. Do you believe this?

Bell: No, but the effect is small. I would really like to know how the microturbulence does vary with T_{eff}, g and [M/H] but until then I think it is satisfactory to use a constant DBV, especially since we are working differentially relative to ϕ^2 Ori when using colours.

Morgan: By 'strong CN' in ω Cen cluster star, do you mean comparable to a Hyades Giant, or stronger?

Bell: The comparison is made relative to other ω Cen giants. Dickens and I haven't compared with spectra of Hyades giants but I guess they would have weaker CN than the strong CN ω Cen stars.

Griffin: I would not trust a quantitative determination of C^{12}/C^{13} based on the C^{13} H spectrum of a K star. Measurements made in the Arcturus Atlas in the region around λ 4300 Å cannot be reliable. Contrary to what you said, my result for C^{12}/C^{13} was based upon CN measurements made in the infra-red region of the spectrum.

Bell: I know you used the C^{13} N lines for your work, as have several people.

Numerous C^{13}H lines can be seen in the spectrum of Arcturus and given a C^{12}/C^{13} ratio consistent with that found from the CN data. This is shown in Krupp's thesis.

Griffin: Have Lambert and Tomkin calibrated their photoelectric technique against high-dispersion photographic spectrophotometry? Their photoelectric equivalent widths published for $C^{12}N$ lines of the (4.0) band in β Gem differ systematically from measurements made photographically at 1.5 Å mm^{-1}, and this suggests that their results may not be quite as accurate as they claim. (you expressed preference for their method, which yields numerical results *quickly*).

Bell: I have not seen their equipment, but the spectra shown by Lambert and Tomkin look very precise and I see no reason to doubt them. I am sure they would be interested in a comparison with your work.

Griffin: I would prefer to make measurements slowly and accurately.

Bell: I am all for accurate measurements.

Griffin: Why did you find it *essential* to adopt a metal abundance in the Hyades giants that is 2 or 3 times (?) the solar abundance? Just because a metal excess would seem desirable to explain broad-band photometry, are you not according undue favour to photometry in the face of spectroscopic evidence?

Bell: Some authors have determined the abundance of μ Leo relative to Hyades giants, others relative to the Sun. A value for the abundances of Hyades giants has to be adopted in order to compare these results. Numerous values exist in the literature and a compromise value of [Fe/H] = 0.3 seems a good one. If the Hyades dwarfs have the same value as the giants then it may be difficult to explain the ultra-violet excess of the Sun relative to the Hyades main sequence with a lower Hyades metal abundance.

Furenlid: We have been concerned with the discrepancy in microturbulence between sun and G dwarfs. We find there is no discrepancy; i.e. if a flux spectrum of the Sun is analyzed by stellar analysis methods the Sun too has a large microturbulence ($\sim$ 3,0 km s^{-1}) if we use the older *gf* values. Microturbulence is a very method-dependent parameter and can normally not be interpreted as a direct measure of physical small scale motions in stellar atmospheres. Care is necessary in choosing the correct value for this parameter in a particular atmosphere analysis, in particular when no weak lines have been measured.

Bell: Maybe the microturbulence is really zero.

Edith Müller: Sorry! For the solar atmosphere we need microturbulence. We have not shuffled the solar microturbulence under the rug.

Bell: You are being very persuasive.

Müller: We have now fairly good model atmospheres for the sun which permit to reproduce very well the centre-limb observations of solar spectral line profiles. But we must in all cases introduce a microturbulence in order to keep the abundance unchanged from centre to limb. This is true for calculations both in LTE and in non-LTE. The microturbulence value near the limb (horizontal component) is larger than the value at the centre of the disc (radial component).

Bell: You're even more persuasive.

Philosophically, I would like a low solar microturbulence (preferably zero microturbulence in fact). However, as long as the Sun is the same as other G dwarfs we should have no problems. It seems from this discussion that the microturbulence at the centre of the disc differs from that of the disc as a whole. We must allow for this fact when using solar curves of growth.

Foy: I have two comments about microturbulent velocity:

(1) The determination of the microturbulent velocity ξ of dwarfs must take into account the splitting of the curve-of-growth (Foy, 1972). I think that large values of ξ in solar type dwarfs are obtained by neglecting this splitting.

(2) I think that the microturbulent velocity is not uniform among G and K giants – I have found (Foy, thesis 1974) that stars near the giant branch have high microturbulent velocities: $\xi \gtrsim$ 1.5 km s^{-1}, whereas horizontal branch giants have small microturbulent velocities: $\xi \lesssim$ 1.0 km s^{-1}.

Bell: I quite agree with your first point and appreciate learning about your second point.

Seitter: You made a strong point in favour of photometry on the basis of the location of a certain star in a two-colour diagram and compared it to the large scatter in the curve-of-growth of a star in M 92. Looking at the error bars in the two-colour diagram it seems that the uncertainty in metal abundance is rather large – how does it compare to the uncertainty of the abundance derived from the curve-of-growth? I ask this question because it seems that the *single* value obtained in photometry simply *suggests* a higher accuracy as one is inclined to overlook the errors while the curve-of-growth always reveals its scatter.

Bell: I think Dr Osborn has a comment on this point. Let me say that the error bars are for Draco, where the stars observed are fainter than in the globular clusters.

Osborn: Let me point out that the quoted error of the high dispersion spectroscopy curve of growth analysis was $\sim \pm 0.7$, which is roughly equivalent to the errors for the metal abundance determination from photometry.

Williams: First, a question: in your calibration of the DDO photometry, how much of your [M/H] measures are CN abundance and how much metallic abundance?

Secondly, a comment on the comparison of the spectroscopic and photometric methods of abundance determination (I have done both and have no axe to grind). The spectroscopic abundances suffer from the problems of the calibration of photographic plates and the subjective placing of the continuum, which are avoided in photoelectric photometric methods. The latter are more suitable for comparing large numbers of stars and demonstrating abundance differences between them.

Bell: In the C (42–45) versus C (45–48) diagram which I showed, the effects of CN line blocking seem quite negligible. This also holds true for changes in the model structure caused by changes in the strengths of the red CN lines. Other DDO indices e.g. C (41–42) are affected by CN, of course.

Querci: (1) What HCNO abundances have you used for computing molecular synthetic spectra?

(2) Have you seen the effect of various HCNO abundances on the synthetic spectra and do you think that you do not get similar synthetic spectra from different model atmospheres?

For example, we have seen that two CN synthetic spectra generated by two model atmospheres different by T_e, g and HCNO are nearly similar, while the CO and C_2 synthetic spectra are never similar for the used parameters.

Bell: For our colour calculations in 1974 (Branch and Bell, *Monthly Notices Roy. Astron. Soc.* (in press), Bell and Gustafsson, 1975, *Dudley Observatory Report* **9**, 319) we used $H = 12.00$, $C = 8.38$, $N = 7.88$ and $O = 8.62$. Before that time, and at the present time, we are using $H = 12.00$, $C = 8.62$, $N = 8.00$ and $O = 8.86$ as the solar values.

Yes, we have run spectra using different CNO abundance ratios. Generally we use spectra in combination with colours (to get temperatures etc.) so hopefully this will minimize the problem of getting similar CN spectra from different models.

ABUNDANCE EFFECTS FOR THE A0–G2 STARS
IN THE GENEVA SYSTEM

B. HAUCK

Institut d'Astronomie de l'Université de Lausanne et
Observatoire de Genève, Switzerland

Abstract. The abundance effects on the tri-dimensional representation for the A0-G2 stars in the Geneva system are reviewed. B2−V1 and d are not affected by the Am characteristic, but B2−V1 is too blue (blocking effect on V1) and d smaller for the Ap stars. For B2−V1 $\geqslant 0.230$, we have a residual effect Δ(B2−V1) $= 1.20$ ($\Delta m_2^* + 0.060$). d is also affected and the residual effect is $\Delta d = -0.4$ Δm_2 for $\Delta m_2 \geqslant -0.060$ and $\Delta d = -1.1$ Δ(B2−V1) -0.024 for $\Delta m_2 < -0.060$.

The abundance effects on the relations between B2−V1 and the parameters of temperature in other systems are studied.

1. Introduction

A tri-dimensional representation for the A0-G2 stars of luminosity classes V to III was presented some years ago (Hauck 1968, 1973a). The following parameters are used:

B2−V1 as temperature parameter
d = (U−B1) − 1.430 (B1−B2) as luminosity parameter
m_2 = (B1−B2) − 0.457 (B2−V1) as blanketing parameter

and [Fe/H] = 6.830 Δm_2 + 0.203   for [Fe/H] > -1.0
± 0.16 ±0.961   ± 0.097

with $\Delta m_2 = m_2$ (star) − m_2 (Hyades)

A review of the properties of the diagram d vs B2−V1 was given by Hauck (1975a). On this occasion an extension to the supergiants was proposed. Using such a diagram it is possible to distinguish the various classes of luminosity.

The colour index B2−V1 is well correlated with Θ_{eff} or with the spectral type. A recent calibration of B2−V1 for the main-sequence stars is given in Table I of Hauck and Magnenat (1975). The values of Θ_{eff} adopted for this calibration are taken respectively in the papers of Schild *et al.* (1971) and Oke and Conti (1966) for the first calibration and in the paper of Morton and Adams (1968) for the second. Calibrations for parameters of temperature in other photometric systems are also given in this table.

In the present paper we have examined abundance effects firstly on B2−V1 and d and secondly on the relations between B2−V1 and the parameters of temperature in other systems.

B. Hauck and P. C. Keenan (eds.), Abundance Effects in Classification, 67–69. *All Rights Reserved.*

2. Abundance Effects on B2–V1

2.1. STARS WITH B2–V1 < 0.230 (A0-F5)

In this range of spectral types, we have the Ap stars and the Am stars. For the latter Hauck and van 't Veer (1970) have shown that no effect was present on B2–V1. On the other hand, for the Ap stars, Gerbaldi *et al.* (1974) have shown that B2–V1 is too blue and that this effect is due to a blocking effect on V1. B2–G is practically not affected by this effect and seems for Ap stars a good parameter of temperature. The following relation for normal stars was derived between B2–V1 and B2–G:

$$B2\text{–}G = 1.386\,(B2\text{–}V1) - 0.437$$

2.2. STARS WITH B2–V1 > 0.230 (F5–G2)

In this range of spectral types, we have a residual effect of blanketing on B2–V1, if Δm_2 ⩽ −0.060. This effect is given by

$$\Delta(B2\text{–}V1) = 1.20\,(\delta m_2 + 0.060)$$

δm_2 is the value determined before the correction on B2–V1, Δm_2 is the value obtained after the blanketing correction on B2–V1 and it is this value which is correlated with [Fe/H].

3. Abundance Effects on *d*

For the Am stars no abundance effects were detected on *d* (Hauck, 1967), while for the Ap stars, Gerbaldi *et al.* (1974) have shown that *d* is affected by the Ap characteristic, in the sense that many Ap stars have a smaller Balmer discontinuity (or a *d* value) than normal stars. This effect is certainly correlated with the blocking effect on V1 (Hauck 1975b). For stars with B2–V1 > 0.230, we have found (Hauck 1973a) that the residual blanketing effect on *d* is

$$\Delta d = -0.4\,\Delta m_2 \text{ for } \Delta m_2 \geqslant -0.060$$
$$\Delta d = -1.1\,\Delta(B2\text{–}V1) - 0.24 \text{ for } \Delta m_2 < -0.060$$

4. Abundance Effects on the Relations between B2–V1 and Temperature Parameters in Other Photometric Systems

Hauck and Magnenat (1975) have given for the main-sequence stars the relations between B2–V1 and B–V, R–I, b–y, Hβ, $(G$–$I)_6$, $(B$–$I)_6$, Y–V and Y–S parameters belonging respectively to *UBVRI*, *uvby*β, 6-colour and Vilnius systems.

In the present study, we have examined whether these relations are also valid for Am, Ap and subdwarf stars. For Am stars we have chosen only those with known spectral types (Hauck, 1973b) and for Ap stars those belonging to the Osawa catalogue (1965). The subdwarfs are principally stars belonging to the lists of Cayrel (1968) and Sandage and Eggen (1959).

The result of our examination is given in Table I. It is important to note that it is only an indication concerning the validity (or the non-validity) of a relation between two colour indices and not an indication whether or not a blanketing effect is present on these indices.

TABLE I

$B2-V1$ vs: for	$B-V$	$R-I$	$b-y$	$B-I$	$G-I$	$Y-V$	$Y-S$
Am	●	●	○	○	●	●	●
Ap	●	○	○	○	○	●	●
Subwarfs	●	●	○	●	●	○	●

● the relation is not valid
○ the relation is valid

References

Cayrel, R.: 1968, *Astrophys. J.* **151**, 1005.
Gerbaldi, M., Hauck, B., and Morguleff, N.: 1974, *Astron. Astrophys.* **30**, 105.
Hauck, B.: 1967, *Publ. Obs. Genève*, No. 74.
Hauck, B.: 1968, *Publ. Obs. Genève*, No. 78.
Hauck, B.: 1973a, in B. Hauck and B. E. Westerlund (eds.) Problems of Calibration of Absolute Magnitudes and Temperature Scales, *IAU Symp.* **54**, 117.
Hauck, B.: 1973b, *Astron. Astrophys. Suppl.* **10**, 385.
Hauck, B.: 1975a, in A. G. D. Philip (ed.), *Multicolor Photometry and the Theoretical HR Diagram* (in press).
Hauck, B.: 1975b in C. Jaschek (ed.), *Physics of Ap Stars, IAU Coll.* No. 32 (in press).
Hauck, B. and Magnenat, P.: 1975, in A. G. D. Philip (ed.), *Multicolor Photometry and the Theoretical HR Diagram* (in press).
Hauck, B. and van 't Veer, Cl.: 1970, *Astron. Astrophys.* **7**, 219.
Oke, J. B. and Conti, P. S.: 1966, *Astrophys. J.* **143**, 134.
Osawa, K.: 1965, *Annals Tokyo Obs.* 2nd ser. **IX**, No. 3.
Sandage, A. R. and Eggen, O. J.: 1969, *Monthly Notices Roy. Astron. Soc.* **119**, 278.
Schild, R., Peterson, D. M., and Oke, J. B.: 1971, *Astrophys. J.* **166**, 95.

DISCUSSION

Osborn: When you speak of 'subdwarfs' what do you consider as a subdwarf? Is it a star with [Fe/H] < -0.5, less than -1.0, or what limit?
Hauck: The more deficient stars! With [Fe/H] < -1.0.

ABUNDANCE EFFECTS ON *uvby* PHOTOMETRY

D. L. CRAWFORD

Kitt Peak National Observatory*, U.S.A.

and

C. L. PERRY

Louisiana State University, U.S.A.

Abstract. Abundance differences from one star to another cause differences in the observed parameters of the Strömgren four-colour system. The major effect is on the m_1 parameter, of course. This paper describes these effects on the parameters.

The *uvby* system was designed by B. Strömgren to separate the surface gravity and abundance effects that appear for F-type stars in a parameter such as $\delta(U-B)$. The parameter δm_1 is a measure of abundance, essentially free of gravity effects, while δc_1 is a gravity parameter, essentially free of abundance effects.

By comparing $(R-I)$, β, and $(b-y)$ for stars with different δm_1 values, we find no correlation with δm_1 for the deviations from an average $(R-I)$ vs β relations. There are small correlations with δm_1 for the residuals from an average $(b-y)$ vs β or $(b-y)$ vs $(R-I)$ relation .

In this paper we derive the δm_1 values with β as the independent parameter. We call the parameter δm_1 (β) to distinguish it from δm_1 $(b-y)$, where $(b-y)$ is the independent parameter.

The relation between δm_1 (β) and [Fe/H] values tabulated by Cayrel and Cayrel de Strobel (1966) is [Fe/H] = 0.20−12 δm_1 for 41 stars, with a standard error in one value of [Fe/H] of ±0.21. The Hyades abundance is 0.20, and the solar δm_1 is +0.^{m}015.

The relation between δm_1 $(b-y)$ and the [Fe/H] values was determined to be [Fe/H] = 0.18−14 δm_1 $(b-y)$ for 49 stars, with a standard deviation in [Fe/H] for one value of ±0.22.

Reference

Cayrel, R. and Cayrel de Strobel, G.: 1966, *Ann. Rev. Astron. Astrophys.* **4**, 1.

* Operated by the Association of Universities for Research in Astronomy, Inc., under contract with the National Science Foundation.

· DISCUSSION

Morgan: What is your current estimate of the magnitude limit of observation where no serious loss in precision is encountered?

    Crawford: About 15th magnitude is reasonable length of time.

    P. E. Nissen: In your $\beta - m_1$ diagram many stars fall below the Hyades relation. Is that due to observational errors of m_1, or do you think that it indicates that there exist several stars more metal rich than the Hyades?

    Crawford: Most are probably due to observational scatter, but not all; some stars appear to be more metal rich than the Hyades.

    Williams: Which is the metal rich ([Fe/H] $\sim$ +0.34) star in your calibration diagram?

    Crawford: HR 3951 (20 L Mi).

A TEST OF THE ACCURACY OF LOW DISPERSION OBJECTIVE PRISM: SPECTRAL CLASSIFICATION OF LATE-TYPE STARS USING DDO PHOTOMETRY

J. J. CLARIÁ* and W. OSBORN

Instituto Venezolano de Astronomía, Mérida, Venezuela

Abstract. A test has been made of the reliability of the multidimensional classification of late-type stars from low dispersion objective prism plates recently attempted by Stock and Wroblewski. Such classification at low dispersion is difficult due to the problem of separating the effects of luminosity from those of abnormal metal abundance. A sample of the stars classified by Stock and Wroblewski as metal weak (pec) and of those classified as luminous stars (class I) were observed using the DDO intermediate-band system. The photometry shows that the stars classified as pec are indeed population II giants, of low metal abundance ($[Fe/H] < -1.0$). The stars classified as I, however, were found in general not to be true supergiants but rather a mixture of various types of giants, such as CN strong stars, with spectral features that resemble, in one way or another, those of higher luminosity stars.

DISCUSSION

Stephenson: At Cleveland, Sanduleak and I have worked with a large amount of spectral plate material obtained with the identical equipment as used by Stock and Wroblewski. We, too, noted many spectra showing the same peculiarities as described by Osborn. Working on a much smaller sample of such stars than he did, but at lower galactic latitude and using slit spectra obtained by me, rather than photometry, we essentially confirm the results quoted by Osborn for such objects.

* Visiting astronomer of Cerro Tololo Inter-American Observatory, operated by the Association of Universities for Research in Astronomy Inc., under contract with the National Science Foundation.

OBSERVED RELATIONS BETWEEN PHYSICAL PROPERTIES AND MK CLASSIFICATION AS FUNCTIONS OF METAL ABUNDANCE FOR F5-K4 STARS

M. GRENON

Observatoire de Genève, Switzerland

Abstract. The Geneva photometric system has been calibrated in terms of [M/H], θ_{eff}, M_v in the spectral range F5 to K4. As the spectral type is a datum generally available, we derive empirical relations showing the coupling of θ_{eff} and [M/H] at given spectral type and luminosity class. Similar relations are offered for the absolute magnitudes and provide a more accurate means for deriving spectroscopic parallaxes. Systematic effects on the estimation of the luminosity class are also shown.

The influence of the chemical composition on the MK classification is examined from the point of view of detecting the possible selection biases arising when star samples are defined by an interval of spectral type (ST) and luminosity class.

In order to estimate these biases the knowledge of two sets of relationships is necessary: the coupling between T_{eff} and [M/H] at given ST and luminosity class, the local dependence between luminosity class and gravity as a function of metal abundance.

As a large number of stars, with known ST and well distributed in metal abundance, is needed to establish these relations, they are deduced from a calibration of the Geneva photometry for the late-type stars. The main properties of this system and the methods of calibrations are described elsewhere (Grenon, 1975a, b). The necessity of a homogenisation of the published lists of [Fe/H] before applying them to statistical investigations is pointed out.

Figure 1 shows the T_{eff} vs [M/H] diagram with the mean loci at constant ST for the dwarf stars; the [M/H] ratio is taken at 0.15 for the Hyades. At given S.T., the range in T_{eff} reaches 700 K, and since the scattering in ST is really small, the internal consistency of the spectroscopic couple of data [Fe/H] and T_{eff} can be checked. From this nomograph can be deduced the minimum interval in ST which must be considered to ensure the completeness in abundances for a stellar sample, in a given range in θ_{eff} or in mass.

A similar diagram M_v vs [M/H] has also been calibrated in ST, for unevolved stars; it can be used for deriving more accurate spectroscopic parallaxes if the abundances are known, since the range in M_v, at constant ST, may be as large as 1.5 mag.

Figure 2 shows the T_{eff} vs [M/H] diagram for the giants. The definition of spectral type appears strongly dependent of [M/H] and most of the samples defined by a given interval of ST lead to biased distributions of metal abundances since we collect objects in differently populated portions of the giant branch. In particular, the choice of ST, later than G8 favours the selection of metal-rich stars. On the other hand the G2-G5 interval appears almost entirely populated by metal-weak and double stars with composite spectra.

B. Hauck and P. C. Keenan (eds.), Abundance Effects in Classification, 75–78. All Rights Reserved.
Copyright © 1976 by the IAU.

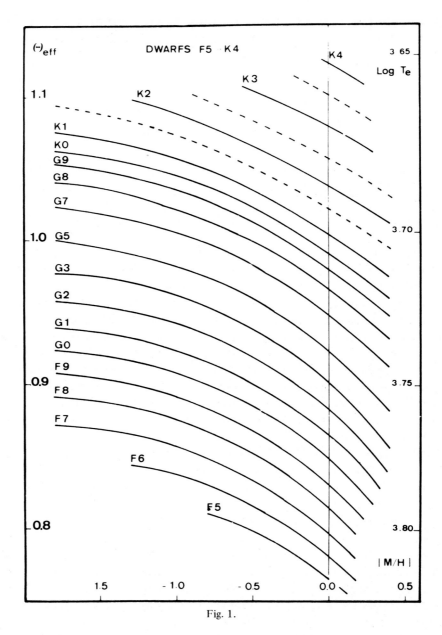

Fig. 1.

The halo giants are generally classified as subgiants or even dwarfs. The tendency to underestimate the luminosity class of metal-poor stars or to overestimate that of the metal-rich ones, is general and valid even for small changes in abundance. At constant gravity this systematic drift can reach one class for a variation of a factor 2 in abundance. In the interval of −1.50 to +0.50 in [M/H], we observe a mean deviation of ± one

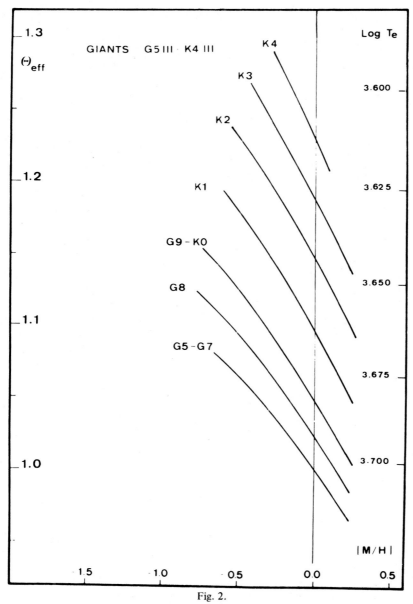

Fig. 2.

luminosity class for a change of ± 0.40 in [M/H]; this tendency suffers many exceptions, nevertheless an interval of completeness in abundances can generally be determined. A selection of stars according to the luminosity class favours metal-rich objects if we choose the class III, rejects some SMR stars if we take the class V. The class IV appears seriously contaminated by metal-deficient giants and binaries formed by a red giant and a main-sequence star.

References

Rufener F.: 1975 'Catalogue', *Astron. Astrophys.* (in preparation).
Grenon M.: 1975a, *Dudley Obs. Report*, 9, 413.
Grenon M.: 1975b, Thesis (in preparation).

DISCUSSION

Cayrel: What is the luminosity MK class you attributed to HD 122563, 165195 and 221170?
Grenon: The second one is not yet analyzed, but for the two others, it is a class II.

NARROW-BAND PHOTOMETRY OF A AND F STARS

E. E. MENDOZA V.

Instituto de Astronomía,
Universidad Nacional Autónoma de México

Abstract. The α and λ photometric systems are used to obtain a bi-dimensional representation of the A and F stars.

The α and λ photometric systems (cf. Mendoza, 1971 and 1975) have been used to measure total absorptions of Hα and O I(λ 7774 Å) lines of A and F stars. The observations were carried out with the 40-in telescope at Tonantzintla, from January to May 1975.

The comparison between Λ and λ and Hβ and Hα photometries is very good. The details will be shown elsewhere.

Provisional results are given in Table I.

TABLE IA

Mean α-indices

Ia	Ib	II	III	IV	IV–V	V	Type
		0.180	0.238	0.225		0.240	A5
				0.249		0.264	A6
0.190			0.230	0.245	0.234	0.248	A7
						0.216	A8
	0.159		0.152		0.195	0.191	F0
					0.162	0.160	F2

TABLE IB

Mean λ-indices

Ia	Ib	II	III	IV	IV–V	V	Type
		0.057	0.038	0.026		0.034	A5
				0.027	0.030	0.028	A6
0.048			0.027	0.027		0.028	A7
						0.021	A8
	0.059		0.036	0.026		0.034	F0
				0.022		0.017	F2

Preliminary conclusions for A and F stars are:

(1) The α-index is a good indicator of stellar gravity for stars of equal spectral type.

(2) The α-index is also a good indicator of stellar atmospheric temperature, for stars of equal luminosity class.

(3) The α-index correlates well with the β-index.

(4) The λ-index isolates very well high luminosity stars (classes I and II) from low luminosity classes.

(5) The λ-index does not, apparently, depend on temperature.

(6) The λ-index, for low luminosity classes, is probably related to an abundance effect.

(7) The λ-index is smaller for metallic line stars than for 'normal' stars.

References

Mendoza, E. E.: 1971, *Bol. Obs. Tonantzintla y Tacubaya*, 6, 137.
Mendoza, E. E.: 1975, *Publ. Astron. Soc. Pacific* 87, 505.

DISCUSSION

Spinrad: What are you planning to do with this new photometric system?

Mendoza: It already has been applied to several kinds of stars with very interesting results. I plan also to observe T-Tauri stars in the near future.

Gerbaldi: In your λ-index can you separate the Am stars?

Mendoza: The classic Am stars are very well separated.

Hauck: Have you examined if a correlation exists between your parameter for the Am stars and the Δm_1?

Mendoza: Not yet, the work reported today is based on recent observations.

Baschek: Should one not expect that the infrared oxygen triplet is sensitive to microturbulence in the range considered, so that this effect should be taken into account besides an abundance effect.

Mendoza: The O I lines at λ7774 Å are primarily sensitive to the gravity of high luminosity stars. Preliminary results indicate that these lines may somewhat depend on chemical composition, stellar age and/or microturbulence, etc. The observations under way will decide on this subject very soon.

De La Reza: Concerning Dr. Baschek's remark I want to say that the oxygen triplet in the near infrared is maybe not so sensible to microturbulence if you consider non-LTE effects. In fact some people have shown that the LTE supersonic microturbulence became non-LTE subsonic microturbulence.

Keenan: In answer to Dr. Baschek, the luminosity effect in O I λ 7744 is very strong through G0, and falls off rapidly after that.

THE HELIUM-TO-HYDROGEN RATIO OF STARS IN YOUNG CLUSTERS AND ASSOCIATIONS

P. E. NISSEN

Institute of Astronomy, University of Aarhus, Denmark

Abstract. A photoelectric narrow-band index, I(4026), of the He I λ 4026 line has been observed with an echelle spectrometer for B-type stars in young clusters and associations.

Model-atmosphere computations of the Strömgren $[c_1]$ index, the Hβ index, and I(4026) as functions of T_{eff}, g and the helium-to-hydrogen ratio have been used to derive these atmospheric parameters from the observed indices for each star.

For the OriOBIb, LacOBI and Sco-Cen associations 11 out of 48 stars in the temperature range from 14 000 to 20 000 K are found to be helium poor in the sense that the derived helium-to-hydrogen ratios lie in the range from 0.05 down to very low values. In the temperature range from 20 000 to 30 000 K no helium poor stars are found, but two stars in OriOBIb are helium rich. Both of these stars show rather strong variations of the I(4026) index.

All stars in h + χ Per and CepOBIII are apparently helium deficient compared to nearby field stars.

In a review paper on photoelectric narrow-band photometry B. Strömgren (1966) proposed to observe an index of one of the helium lines in the spectra of B-type stars in order to determine the helium abundances of a large sample of stars in the Galaxy. Following this proposal I have observed an index, $I(4026)$, of the strength of the He I λ 4026 absorption line for several hundred B-type stars.

The $I(4026)$ index is defined as:

$$I(4026) = 2.5 \log \frac{F_{14}(4026)}{F_6(4016) + F_6(4036)} ,$$

where $F_{14}(4026)$ is the flux of the star in a 14 Å wide spectral band centred on He I λ 4026, and $F_6(4016) + F_6(4036)$ is the total flux of two 6 Å wide 'continuum' bands at 4016 Å and 4036 Å respectively. The bands are defined by exit slots in the spectrum formed by an echelle spectrometer with a linear dispersion of 1.3 Å mm^{-1} at 4026 Å. The intensity of the light passing the slots is measured with pulse-counting technique.

In an earlier publication (Nissen, 1974) it was described how model-atmosphere computations of the Strömgren $[c_1]$ index, the Hβ index, and $I(4026)$ as functions of T_{eff}, g and the helium-to-hydrogen ratio, $\epsilon(\text{He})=N_{\text{He}}/N_{\text{H}}$, could be used to derive these stellar atmospheric parameters for each observed star. It was found that the mean helium abundance of field stars, that lie within 500pc from the sun, was about ϵ (He) = 0.10. The rms scatter of ϵ(He) was found to be smaller than 0.01. Furthermore stars in the rather distant cluster, NGC 6231, were found to have about the same mean helium abundance as the field stars.

In the present paper I want to describe some new results, that are based on observations of $I(4026)$ recently obtained with the ESO 100 cm telescope at La Silla, Chile, and the 193 cm telescope at Observatoire de Haute Provence, France. The observations

B. Hauck and P. C. Keenan (eds.), Abundance Effects in Classification, 81–83. *All Rights Reserved*

include the Sco-Cen, OriOBI$_b$ and LacOBI associations, for which the whole range of B spectral types was covered, and $h + \chi$ Per and CepOBIII, for which the observations were confined to the spectral range B0-B3.

For Sco-Cen, OriOBI$_b$ and LacOBI 11 out of 48 stars in the effective temperature range from 14 000 to 20 000 K are found to be helium poor in the sense that the derived helium abundances lie in the range from $\epsilon(He) = 0.05$ down to very low values. In Sco-Cen these stars are HD 142884, 144334, 151346, 142990, 143699, and 146001, of which the first three have been classified as weak helium-line stars by Garrison (1967). In OriOBI$_b$ the helium-poor stars are HD 36046, 36526, 37525, 37642, and finally one star in LacOBI, HD 213918, was found to be helium poor. For all these stars there is a rather large inconsistency between the $[c_1]$ index and the spectral type, but I remark that for the stars in OriOBI$_b$ and LacOBI I have been able to find HD types only.

In the effective temperature range from 20 000 to 30 000 K no helium-poor stars were found, but two stars in OriOBI$_b$ are helium rich. These stars are σOriE and HD 37776. Both are found to be variable in $I(4026)$, and for σOriE Thomsen (1974) has found the variations to be periodic.

The spectral range B0-B2 is particularly suitable in deriving the helium abundances, because the model-atmosphere computations have revealed that a linear combination of $I(4026)$ and the Hβ index, that is observed by D. Crawford and collaborators, is nearly independent of effective temperature and surface gravity, but a sensitive function of the helium-to-hydrogen ratio. This means that differential helium abundances can be directly determined from the observed values of $I(4026)$ and Hβ, both of which indices are independent of interstellar reddening.

The mean helium abundances for different groups of stars as derived from stars in the spectral range B0-B2 are given in Table I. N is the number of stars in each group, and $\langle y \rangle$ is the mean logarithmic helium-to-hydrogen ratio. The $\langle y \rangle$ value for the field stars is normalized to -1.00, and the quoted errors, that are derived from the rms scatter of the individual y values, refer to differential helium abundances. From Table I I conclude that two groups of stars, namely $h + \chi$ Per and CepOBIII, have an apparently lower helium abundance than the rest of the stars observed.

Details of this work will be given elsewhere.

TABLE I

Group	N	$\langle\epsilon(He)\rangle$	$\langle y \rangle \pm \sigma(\langle y \rangle)$
Field stars	33	0.100	-1.00
Sco-Cen	12	0.101	-0.99 ± 0.02
NGC 6231	10	0.087	-1.06 ± 0.03
$h + \chi$ Per	12	0.059	-1.23 ± 0.03
CepOBIII	10	0.060	-1.22 ± 0.03
LacOBI	6	0.085	-1.07 ± 0.03
OriOBI$_b$	8	0.091	-1.04 ± 0.03

References

Garrison, R. F.: 1967, *Astrophys. J.* 147, 1003.
Nissen, P. E.: 1974, *Astron. Astrophys.* 36, 57.
Strömgren, B.: 1966, *Ann. Rev. Astron. Astrophys.* 4, 433.
Thomsen, B.: 1974, *Astron. Astrophys.* 35, 479.

DISCUSSION

Spinrad: Please elaborate on the galactocentric distances of the clusters with various He abundances, which you have just described.

Nissen: The associations with a helium-to-hydrogen ratio of $N_{He}/N_H = 0.10$, namely Sco-Cen, Ori OBI$_b$, Lac OBI$_b$, all lie within a distance of 600 pc from the Sun. The NGC 6231 cluster, for which $N_{He}/N_H \cong 0.10$, lies in the Sagittarius spiral arm, 2000 pc away from the Sun. The $h + \chi$ Per cluster with the low helium abundance of $N_{He}/N_H = 0.06$ lies in the outer spiral arm at a distance of 2000pc. Finally CepOBIII, for which $N_{He}/N_H = 0.06$ lies at a distance of 800 pc in nearly the same direction as $h + \chi$ Per.

Stephenson: Did you say anything about the non-influence of stellar rotation on your 4026 index, and if so would you please repeat what it was you said?

Nissen: No, I did not say anything about the influence of stellar rotation on $I(4026)$. However, in an earlier publication (*Astron. Astrophys.* 36 (1974), 57) I investigated if there was any dependence of the derived helium-to-hydrogen ratios of field stars on rotational velocity and I did not find any effect for projected rotational velocities below 300 km s^{-1}.

Morgan: There is a large systematic error in the Henry Draper Catalogue for faint B stars in the Orion region. Stars having HD types of B8 and B9 have MK types of around B3 to B5.

Nissen: This means that one needs MK types in order to see if there is a discrepancy between colours and spectral types for the stars in Orion, which I find to be apparently helium poor.

Walborn: (1) Can you say anything about periodicities in σ Ori E in your own λ 4026 data?

(2) How large a telescope would be needed to reach stars in the Magellanic Clouds ($\sim 16^m$) with your system?

Nissen: (1) No, I can only refer to B. Thomsen's observations.

(2) With a 4-m telescope it should be possible to determine the $I(4026)$ index for the early B-type stars with an accuracy of $0.^m005$ if several hours of integration time per star is applied.

RECENT RESULTS PERTAINING TO THE HELIUM-RICH STARS

N. R. WALBORN

Cerro Tololo Inter-American Observatory, La Serena, Chile

Abstract. The term *helium-rich stars* here refers to those stars with apparent spectral types near B2 and helium lines which are enhanced, but not stronger than the hydrogen lines. The previously known helium-rich star HD 184927 has been found to be a spectrum variable; the equivalent widths of the helium lines changed by an average of 46% between two observations. HD 186205 has been newly recognized as a member of the helium-rich class. Further work toward understanding the rapidly variable Hα emission in the prototype helium-rich star, Sigma Orionis E, is in progress in co-operation with Dr J. E. Hesser at CTIO.

DISCUSSION

Garrison: I have spectra of stars in the small tight cluster which includes the multiple star σ Ori. The cluster forms an extremely narrow main sequence and contains a 'He weak' star in addition to σ Ori E, the 'He rich' star. The 'He weak' star is a mild one with a B5 colour and a B8 spectral type.

Morgan: On the assumption that σ Ori E is a member of the Orion association, do you find that it has normal luminosity for a B2V star?

Walborn: The presence of σ Ori E in a multiple system which contains an earlier-type main-sequence star, as well as a normal B2V star with nearly identical apparent magnitude and colour to *E*, places strong constraints of interpretations of the helium-rich star.

Bell: Is there any evidence for the existence of He^3 in these stars?

Walborn: No.

Schatzman (To Dr Nissen and Dr Walborn): Mrs S. Vauclair has a nice explanation of the difference between He poor and He rich stars, as a result of competition between diffusion and stellar wind.

Nissen: Yes, I agree that the explanation by S. Vauclair is very interesting.

Note added in proof. A periodicity of $1^{d}19$ has been found in the spectrum, light, and color variations of Sigma Orionis E. References: *Publ. Astron. Soc. Pacific* **87**, 613, 1975; *Astrophys. J. Letters*, April 15, 1976.

ABUNDANCE EFFECTS IN THE CLASSIFICATION OF K STARS

P. M. WILLIAMS

Royal Observatory, Edinburgh, United Kingdom

Abstract. The influence of metal abundance and gravity on the relation between spectral type and effective temperatures of late G and K type stars is investigated and calibrated using metal abundances from narrow-band photometry, near infrared photometry and independent luminosity estimates.

1. Introduction

While spectral classification of the coolest and most highly evolved stars has always reflected their composition owing to the conspicuous effects on their spectra of the carbon-to-oxygen abundance ratio, the effects of the overall metal abundance on stars' spectra are more subtle. They are also more liable to misinterpretation among those spectral types dependent on metal to hydrogen line ratios and the present contribution is an attempt to calibrate such affects among the G and K type giants. In principle, there are two ways of going about this — either by simulating the classification criteria using synthetic spectra and finding the model atmosphere parameters corresponding to the class boundaries, or by using the stars themselves to investigate the transformation from spectral classifications to the $[T, g, x]$ (Temperature, gravity, metal abundance) space. As the first approach also involves simulation of the application of the classification criteria, the second approach using stars already classified by experienced observers is used here. The G and K type stars cover wide ranges of mass, age and metal abundance and have also been the subjects of numerous abundance analyses, making them very suitable for such a study. It is appropriate at this meeting to recall that nearly 40 years ago, Dr. W. W. Morgan, in a paper (Morgan, 1937) foreshadowing the two-dimensional spectral classification, emphasised the fundamental nature of the effective temperature — surface gravity diagram and its advantages over the HR diagram. The fundamental nature of surface gravity was noted by Pannekoek (1922) who wrote '.... the physical quantity, directly given by the spectra used for the determination of spectroscopic parallaxes, is the gravitation at the surface of the star.' Inevitably, the determination of spectroscopic absolute magnitudes and the calibration of luminosity classifications in terms of absolute magnitude must also involve knowledge of the gravity-luminosity relation which itself requires a mass-luminosity relation. Even if these are known, it can be shown (e.g. Williams, 1971) that a relative error of a factor of two in the gravity, which is by no means pessimistic even for a high dispersion spectroscopic analysis, corresponds to one magnitude in M_v. Therefore gravity is preferable to luminosity or absolute magnitude for the second axis of calibration.

B. Hauck and P. C. Keenan (eds.), Abundance Effects in Classification, 87–90. All Rights Reserved.
Copyright © 1976 by the IAU.

2. Calibration

The stars used for this calibration comprise a sample of G8 to K4 giants, omitting known spectroscopic binaries and stars with 'peculiar' classifications, with spectral classifications by Johnson and Morgan (1953) or Roman (1952) and for which metal abundances were derived in a narrow-band photometric programme. Since this has already been described elsewhere (Williams, 1971, 1972), it is sufficient here to recall that the metal abundances were derived from spectrophotometric indices analysed by spectral syntheses and are accurate to within 0.2 in [Fe/H] and that the temperatures, determined primarily from $(I-K)$ photometry calibrated with model atmospheres and empirically calibrated $(R-I)$, are essentially independent of metal abundance. The temperatures are effectively model atmosphere labels rather than true effective temperatures which cannot be determined experimentally for any but a small number of stars. The gravities were determined from $M_v(K)$ with an abundance term, trigonometric and dynamical parallaxes and theoretical mass-luminosity relations. They are probably accurate to within a factor of two and no calibration of luminosity class against gravity is attempted here.

First, consider the disposition of stars of a given spectral type in $[T, g, x]$ space. This is illustrated in Figure 1 for the K0 stars, which comprise the most populous class. The stars were divided into ranges of width Δ [Fe/H] = 0.2 and $\Delta \log g = 0.3$ and the average

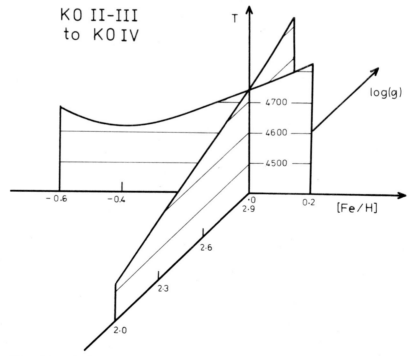

Fig. 1. Disposition of K0 giants and subgiants in temperature-gravity-metal abundance space.

temperature found for the stars in each. Owing to the paucity of stars in much of the field, only those in the gravity range $2.8 \leqslant \log g < 3.1$ were plotted against metal abundance, and stars having $-0.3 < [Fe/H] < 0.3$ were combined for the gravity axis. The features are as one might expect; the only surprise being the turn up in temperature at the low abundance end. This is also observed among the K3 stars, although there are so few of these to make the result even more uncertain, and the G8 stars. Taken together, these results suggest that it is mainly among the mildly metal poor stars, with $[Fe/H] \sim -0.5$, that the classification is most affected by metal abundance. The magnitude of the effect is about 100 K or 1–2 sub-types in spectral class. The dispersion in temperature of stars in any one range is usually between 50 K and 100 K, so that even where the metal abundance and gravity are known, the spectral type is barely a good measure of the temperature. Among stars of later types, the dispersions are generally less.

Secondly, for an overall idea of the sensitivity of the relation between spectral type and temperature, regression lines were fitted to the $T - \log g$ and $T - [Fe/H]$ data for stars of each type, limiting the abundances in the first case to $-0.2 < [Fe/H] < 0.2$ and the gravities in the second by considering stars of luminosity class III only. The slopes of these lines are given in Table I together with the differences $\Delta \log g$ and $\Delta [Fe/H]$ which will affect the spectral classification by one sub type. As there were so few K1 and K4 type stars, the results for these were combined with those for the K2 and K3 stars in the table, although the slopes were, of course, determined separately.

TABLE I

Abundance and gravity effects on classification

Type	$\partial T / \partial \log g$	$\partial T / \partial [Fe/H]$	$\Delta \log g$	$\Delta [Fe/H]$
G8	150	258	0.46	0.27
K0	200	232	0.52	0.45
K1-2	203	245	0.68	0.56
K3-4	168	277	0.79	0.48

The lack of large enough samples of stars with homogeneously determined metal abundances makes it difficult to extend this analysis to stars of other spectral types although it is observed that G-K subgiants behave similarly in the abundance-temperature plane if one includes those analysed by Glebocki (1972) with those from the narrow-band work.

3. Conclusion

The sensitivity to gravity and metal abundance of the relation between temperature and spectral types of G8 to K4 giants has been investigated, indicating that a factor of three in the gravity or two in the metal abundance affects the relation by up to 100 K or about one spectral sub-type.

References

Głebocki, R.: 1972, *Acta Astron.* **22**, 141.
Johnson, H. L. and Morgan, W. W.: 1953, *Astrophys. J.* **117**, 313.
Morgan, W. W.: 1937, *Astrophys J.* **85**, 380.
Pannekoek, A.: 1922, *Bull. Astron. Inst. Neth.* **1**, 107.
Roman, N. G.: 1952, *Astrophys. J.* **116**, 122.
Williams, P. M.: 1971, *Monthly Notices Roy. Astron. Soc.* **153**, 171.
Williams, P. M.: 1972, *Monthly Notices Roy. Astron. Soc.* **158**, 361.

DISCUSSION

Bidelman: What were the sources of your spectral types?
  Williams: Preferably from Johnson and Morgan, otherwise from Roman.

COMPARISON OF COLOUR CURVES OF
MIRA STARS OF SPECTRUM M AND S

C. PAYNE-GAPOSCHKIN

Center for Astrophysics, Cambridge, U.S.A.

Abstract. Colour variations for Mira stars of spectrum Me and Se are deduced from the two micron survey of Neugebauer and Leighton and concurrent visual observations of the same stars.

The colours change cyclically during the periodic change of brightness. Colour is a function of maximal spectrum and also of period. Mira stars of class Se are bluer than those of Me of similar period.

A study of the light and colour curves of Mira stars, based on the infra-red catalogue of Neugebauer and Leighton and concurrent observations compiled by the American Association of Variable Star Observers, is being carried out by the writer and Charles A. Whitney. Neugebauer and Leighton tabulate magnitudes that they designate I and K, corresponding to effective wavelengths about $0.8\,\mu$ and $2.2\,\mu$ respectively. Their K magnitudes correspond to the K magnitudes of the standard Johnson system; their I magnitudes, which correspond to a shorter effective wavelength than Johnson's, (and which we accordingly designate I^1), are related to the Johnson system by the relation:

$$(I-K) = 0.745\,(I'-K) - 0.13.$$

The I and K magnitudes, and the visual magnitudes, which we designate V, although they are not exactly equivalent to V magnitudes, permit us to determine three colour indices, $(I'-K)$, $(V-I')$, and $(V-K)$ for each dated observation.

Colour indices and phases (referred to visual maximum) are available for 125 Mira stars of spectrum M, and for 17 stars of spectrum S, and these form the basis of the present comparison. All the colours vary with phase, being bluest at visual maximum, reddest near visual minimum. The M stars are treated in three period groups (mean periods 239, 350 and 456 days); the mean period for S stars is 380 days.

Figure 1 shows colour-colour curves for the Mira stars and compares them with the colour-colour relation for non-variable stars of luminosity class III. Data for the three period groups are plotted on shifted scales, as described in the legend. On the fall from visual maximum, or slightly later, the points conform to the relation shown by the non-variable stars of the same spectral class. The colours describe counter-clockwise cycles, compatible with I' being relatively brighter on the rising curve than on the fall. When period is taken into account, it is evident that the loop for the S stars makes them somewhat bluer than the corresponding M stars.

A synthesis of the K curves for multiply observed Me and Se stars shows that the K magnitude comes to maximum about 0^{P} *after* visual maximum, an effect also observed

B. Hauck and P. C. Keenan (eds.), Abundance Effects in Classification, 91—93. All Rights Reserved.

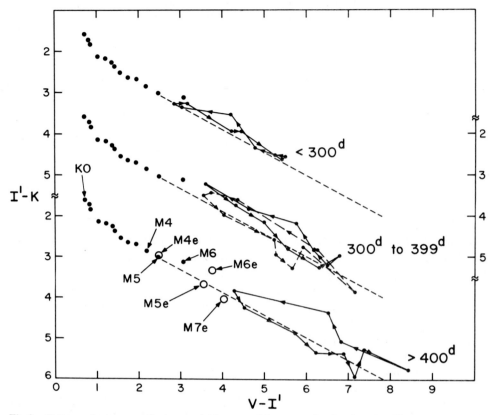

Fig 1. Colour-colour curves for non-variable stars of late type, luminosity class III, and Mira stars of spectrum M and S. Ordinate, in magnitudes, $(I'-K)$; abscissa, in magnitudes, $(V-I')$. Heavy dots (plotted three times on shifted scales of ordinates) are non-variable stars from K0 to M6. Smaller dots, loops described during the cycle of variation for stars of average period 239, 350, and 456 days, class Me. Broken lines, Se stars, mean period 380 days. Dotted lines extend the relation for non-variable stars through points corresponding to the fall from maximum. On the lowest curve, points corresponding to K0III, M4III, M5III and M6III are identified. Circles show the relation between maximal colours for Mira stars of classes M4e, M5e, M6e and M7e.

in the bolometric light curves determined by Pettit and Nicholson. The range of K is somewhat greater for Se than for Me stars.

It is evident that the I' magnitude (and perhaps also the V magnitude) is relatively brighter on the rise from minimum to bolometric (and K) maximum. Whether this excess is related to the development of the bright-line spectrum can only be decided by spectrophotometry.

Interpretation of the colours in terms of temperature requires an independent calibration. Attempts to deduce effective temperatures by means of angular diameters have led to inconsistent results when our data are compared for normal giants, supergiants, and

Mira Ceti. Attempts to derive colour temperatures are vitiated by unknown contributions of band absorption. The fact that S stars are bluer in all three colour systems than M stars of the same period must be studied in this context.

DISCUSSION

Mendoza: How do infrared excesses, shown by some Me and Se stars, affect your results?

Payne-Gaposchkin: I believe that the fact that the Mira stars of all periods are bluer on the rise than on the fall of the light curve, is related to the greater strength of the bright line (and probably the continuum) of hydrogen as the brightness increases. A quantitative answer would require spectro-photometry.

THE COLOUR EXCESS SCALF AND INTRINSIC COLOUR PROPERTIES OF THE LONGER PERIOD CEPHEIDS

R. CANAVAGGIA

Observatoire de Paris, France

and

P. MIANES and J. ROUSSEAU

Observatoire de Lyon, France

Abstract. Following a discussion of two recent papers by Schmidt (1973, 1975), we conclude that for the large amplitude, longer period Cepheids (log P greater than about 1.1), the small colour excesses derived from the six-colour photometry (traditional method or Parsons method) are to be preferred to the large colour excesses derived from the Γ-index photometry (Kraft's scale).

Some papers by E. G. Schmidt were an incitement to clarify the question of the colour excess scale and intrinsic colour properties of the longer period Cepheids (periods longer than about ten days).

We first draw attention to the two following facts:

(1) For these Cepheids, since about 1960, the choice has been between the colour excesses derived from the six-colour photometry (the Lick system), and those derived from Kraft's Γ-photometry – the second ones being larger than the first ones by 0.15 or 0,20 m.

(2) The colour excesses of the six-colour photometry (traditional method or Parsons method) are meant to ensure (on the whole) uniformity and continuity of intrinsic colour properties for all colours (other than colour U), over the entire period range of the long period Cepheids. It follows that Kraft's scale necessarily implies colour discrepancies when going from the shorter to the longer period Cepheids.

Hence the colour and temperature discrepancies that were encountered by Schmidt (1973), when assuming Kraft's scale.

Now, against Kraft's results concerning the longer period Cepheids, three kinds of arguments can be advanced:

– *Arguments against Kraft's method.* There are indications in the work of Nikolov and of Tsarevsky and Yakimova that the Γ spectral type system may not be well defined for the later spectral types.

– *Arguments in favour of the six-colour method.* The basic principle of the method must be recalled (opposite curvatures of the interstellar and temperature reddening laws on either side of colour G).

With this in mind, it is difficult to accept as real the large colour discrepancies between shorter and longer period Cepheids which result from Kraft's scale, as they bear all the characteristic features of differential interstellar reddening.

— *Schmidt's argument* — A third argument was recently provided by Schmidt himself (1975).

Having measured on the *uvbyβ* system the interstellar reddening of field stars in the direction of two longer period Cepheids, Schmidt found that the colour excesses that he formerly assumed for these stars had been largely overestimated.

This last (extrinsic) argument will probably be felt as the more convincing and the one that settles the question: for large amplitude, longer period Cepheids, the small colour excesses of the six-colour photometry are to be preferred to the larger colour excesses of Kraft's scale.

Concerning the intrinsic colour properties of these stars, however, Schmidt's position, in his communication of 1975, is not entirely clarified: He does not seem to realize that to give up Kraft's large colour excesses also means giving up the large colour discrepancies (and line blanketing discrepancies) previously suggested by him.

References

Schmidt, E. G.: 1973, *Monthly Notices Roy. Astron. Soc.* **163**, 67.
Schmidt, E. G.: 1975, *Monthly Notices Roy. Astron. Soc.* **170**, 39P.

DISCUSSION

Cayrel: You said that an overestimated colour-excess 'simulates' a line-blanketing excess in the blue colours. What amount of metal overabundance would be necessary to account for this line-blanketing excess?

Canavaggia: What I meant is that an overestimated de-reddening simulates a line-blanketing excess for E. G. Schmidt, who uses Johnson's system, because it results in associating an $R-I$ index that is 'too blue' with a $B-V$ index that is 'too red'. But when the $IRGBV$ colours of the Lick system are considered, things are different: the over-de-reddening results in a deficiency of I and of V with respect to the median R, G, B colours, and this cannot be accounted for by any metal overabundance.

Spinrad: Will the new colour excesses change the Cepheid distances and then propagate into uncertainties in the extra-galactic distance scale?

Canavaggia: The Period-luminosity and Period-colour-luminosity relations for Cepheids are mainly calibrated through using shorter period cluster Cepheids, and nothing is changed for these stars. Only the extension of the laws towards cooler, more luminous stars depends on longer period Cepheids, and especially on RS Puppis, which Westerlund showed to be an association Cepheid. Sandage and Tammann used this star, assuming the colour excess of the two nearer stars in the association. This excess is about midway between the colour excesses implied by Kraft's scale and the six-colour scale — so half of the change suggested here was already applied by Sandage.

ON THE DERIVATION OF ABUNDANCES BY MEANS OF FEATURES IN THE SEQUENCES OF THE OLD OPEN STAR CLUSTERS

A. MAEDER

Observatoire de Genève, Switzerland

Abstract. Stellar evolution near the main-sequence is still subject to many discussions. In addition to the case of the cluster M 67 (Racine, 1971), quite systematic differences have been encountered during the comparison of theoretical isochrones and observed sequences in the colour-magnitude diagram of the old open star clusters (Maeder, 1974). In the case of M 67 and NGC 188, many attempts to discuss the characteristics of their sequences (such as the gap corresponding to the hydrogen-exhaustion phase) in terms of chemical composition have been made (e.g. Aizenman *et al.*, 1969; Demarque and Schlesinger, 1969; Demarque and Heasley, 1971; Torres-Peimbert, 1971; Hejlesen *et al.*, 1972; Caloi *et al.*, 1974); some of these works have suggested to explain the gap parameters by a high metal content. Attemps to explain the differences between models and observations by means of simplified models with overshooting from convective cores have also been made (Maeder, 1973; Prather and Demarque, 1974). Let us also note that it has been shown that it is unlikely that the anomalies found are due to systematic effects, like the interstellar reddening, binarity, rotation or effects in calibrations or composition of the initial homogeneous models.

We now turn to the interior models, which are challenged by the above anomalies. It is clear that the evolutionary tracks depend closely on the X-profiles in the stellar interiors; these profiles are determined themselves mainly by the size of the convective cores during the evolution. In the usual sets of models of stellar interiors, convection in the cores is not treated explicitly. There are well-known reasons for that (Biermann, 1932), which are certainly fully justified as far as the course of the various structural variables is concerned, but certainly not as far as the extent of the convective zone is concerned. Thus, a new method has been developed (Maeder, 1975) allowing to perform an explicit treatment of convection in the cores with non-local expressions of the mixing-length formalism. An iterative process is used to integrate simultaneously the equations of stellar structure and the equations describing convection.

The application of the method to stars in the range of 1.25 to 3 $M_\odot$ shows that the zone fully mixed by turbulent convective motions extends appreciably farther from the limit given by Schwarzschild's criterion. The overshooting, which amounts to about 14% of a mixing-length for the homogeneous initial models, brings only negligible modifications in the results of the initial models. Thus, the location of the zero-age sequence in the HRD remains unmodified by the overshooting from the cores. However, the consequences of this effect are much larger during evolution and several changes occur in the

B. Hauck and P. C. Keenan (eds.), Abundance Effects in Classification, 97–99. All Rights Reserved.

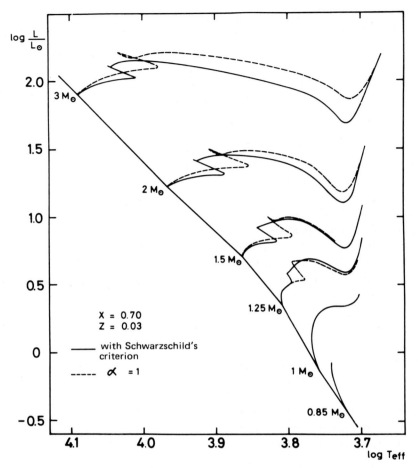

Fig. 1. Evolutionary paths in the theoretical HRD for stars with local theories of convection in the core (Schwarzschild's criterion) and with non-local theories predicting an overshooting from the core. The case, where $\alpha = l/H_p = 1$ in the core is shown. The case $\alpha = \frac{1}{2}$ has also been computed. The overshooting also modifies the lifetimes appreciably.

shape of the tracks and in the lifetimes. Figure 1 shows some results; the case where $\alpha = l/H_p = 1$ in the core is illustrated. The case $\alpha = \frac{1}{2}$ has also been computed. In models with overshooting, the hook in the evolutionary tracks is described very rapidly, so that such models predict that the visible gap will occur almost vertically in the HRD. Actual comparisons show that models with a mild overshooting ($\alpha = \frac{1}{2}$, i.e. the zone of over-shooting contains about 2–3% of M_r/M at $1.5\,M_\odot$) provides simultaneously for all features of the sequences in the HRD a very satisfactory agreement, which is never obtained with the usual grid of models.

From the work, very briefly outlined above, we may make the following remarks concerning the determinations of chemical abundances by means of parameters of cluster sequences: (1). It is necessary to consider in the comparisons of models and observations all the features describing the sequence of a cluster in the region of the gap and not only to consider one or two gap parameters. (2). The case of M 67 and NGC 188 cannot be disconnected from the fact that also for other slightly younger clusters, a very poor agreement is realized between observations and models. (3). The determination of X, Y and Z content by means of gap parameters is rather hazardous, as these parameters are very sensitive to the process of overshooting which is hydrodynamically possible at the edge of the core. (4). More reliance must certainly be given to the determinations of helium and metal content, which are based on the location of the zero-age sequence, (*cf.* Morton and Adams, 1968), as this location is very independent on the extent of the fully mixed zone in the centres of upper main-sequence stars.

References

Aizenman, M. L., Demarque, P., and Miller, R. H.: 1969, *Astrophys. J.* **155**, 973.
Biermann, L.: 1932, *Z. Astrophys.* **5**, 117.
Caloi, V., Castellani, V., and Di Paolo, N.: 1974, *Astron. Astrophys.* **30**, 349.
Demarque, P., Heasley, J. N.: 1971, *Astrophys. J.* **163**, 547.
Demarque, P., Schlesinger, B. M.: 1969, *Astrophys. J.* **155**, 965.
Hejlesen, P. M., Jorgensen, H. E., Petersen, J. O., and Romcke, L.: 1972, in G. Cayrel de Strobel and A. M. Delplace (eds.), *IAU Colloquium.* **17**, Observatoire de Paris.
Maeder, A.: 1973, in Stellar Instability and Evolution, *IAU Symposium No 59*, p. 109.
Maeder, A.: 1974, *Astron. Astrophys.* **32**, 177.
Maeder, A.: 1975, *Astron. Astrophys.* **40**, 303.
Morton, D. C., Adams, T. F.: 1968, *Astrophys. J.* **151**, 611.
Prather, M. J., Demarque, P.: 1974, *Astrophys. J.* **193**, 109.
Racine, R.: 1971, *Astrophys. J.* **168**, 393.
Torres-Peimbert, S.: 1971, *Bol. Obs. Tonantzintla Tacubaya* **6**, 3.

DISCUSSION

Spinrad: Do your new 'overshoot' convective core models change the ages of the old galactic clusters?

   Maeder: The overshooting from the convective core mainly redistributes the lifetimes among the various evolutionary phases. However, the age estimates will probably be increased by about 20%.

   Bell: Could you tell us how much an isochrone will differ, if it is computed with and without allowing for overshooting?

   Maeder: The differences in the isochrone produced by the overshooting depend on the part of the isochrone you are considering. For example, a very large effect exists for the height of the top of the gap above the zero-age sequence. This feature may differ by about $0\overset{m}{.}8$, when overshooting is included. Quantities, usually called ΔM_{bol} (gap) and ΔM_{bol} (peak) are also modified by the overshooting from convective core.

A METHOD FOR DETERMINING THE CHEMICAL COMPOSITION PARAMETERS (X, Y, Z) OF GALACTIC CLUSTERS

WAYNE OSBORN and JUAN J. CLARIÁ*

Instituto Venezolano de Astronomía, Mérida, Venezuela

Abstract. A method is described by means of which the chemical composition parameters (X, Y, Z) may be derived for galactic clusters that contain evolved red giant members and for which the cluster reddening and distance moduli are known. The method is based on observations of the evolved stars utilizing the DDO intermediate-band system. First, the DDO photometry by itself allows one to separate with a high degree of certainty the physical members of the clusters from the field stars. In cases where membership in the cluster can be confirmed, the unreddened DDO colours can be used to obtain the effective temperatures, surface gravities and metal abundances (Z) of the stars. These data combined with the absolute magnitudes available from the distance modulus permit the masses to be calculated. Finally, if some of the observed red giants form part of a clump on the giant brach in the HR diagram, the helium abundance (Y) of the clusters can be estimated by comparing the derived physical parameters of these stars with theoretical models. An example of the application of the method to the galactic cluster NGC 2420 is given.

1. Introduction

In a recent study Gross (1973) showed that the position of a zero age horizontal branch (ZAHB) star in the effective temperature-surface gravity plane is determined almost entirely by its helium abundance. This provides a simple method for estimating the helium abundances of several types of objects, for example the globular clusters. Applying this theoretical result to the small amount of observational data available Gross found the unexpected result that population I objects appear to have smaller helium abundances than population II stars. A compilation of all published helium abundance determinations seemed to support this conclusion, but it was obvious that more and better data would be needed for a definitive answer.

One obvious way to test Gross' results is to compare the results from applying his helium abundance method to globular cluster horizontal branch stars and to their population I analogues the clump stars of evolved galactic clusters. A number of determinations of temperatures and gravities for globular cluster horizontal branch stars have been published (Newell *et al.*, 1969a, 1969b; Philip, 1972; Osborn, 1973) but little data is available for clump stars. It was therefore felt it would be worthwhile to attempt to derive temperatures and gravities for clump stars in evolved galactic clusters. Rather than present definitive results, the object of this paper is to show what has been learned from

---

* Visiting astronomer of Cerro Tololo Inter-American Observatory, operated by the Association of Universities for Research in Astronomy, Inc., under contract with the National Science Foundation.

B. Hauck and P. C. Keenan (eds.), Abundance Effects in Classification, 101–109. *All Rights Reserved.*

some exploratory steps in this direction using the DDO intermediate-band system (McClure and van den Bergh, 1968; McClure, 1973). In particular, we will describe a procedure that can be used to determine the chemical composition parameters of clusters with clump stars for which the reddening and distance modulus are known.

2. Identification of the Clump Stars

The first problem that one encounters is that of separating red field stars from the physical members of the clusters. The best method for doing this is probably on the basis of proper motions and/or radial velocities, but frequently these data are unreliable or not available. We have found that membership can usually be assigned with fair certainty using only UBV and DDO photometry. The DDO $C(45-48) - C(42-45)$ and $C(41-42) - C(42-45)$ diagrams usually allow a decision to be made on the basis of the photometric spectral type and the distance modulus obtained from the corresponding M_v (Janes, 1975). For example, foreground dwarfs usually are easily identified as shown in Figure 1. In regions of heavy interstellar reddening the situation can become more complicated. In such cases, however, the reddening itself can be used as a criterion. Assuming the Whitford reddening curve, McClure and Racine (1969) have derived the relation valid for giants with $0.80 \leqslant (B-V)_0 \leqslant 1.55$:

$$E(B-V) = 2.175(B-V) - 2.380\ C(45-48) - 1.420\ C(42-45) + 1.841$$

where all the photometric indices are the observed (reddened) values. Figure 2 shows a comparison for several clusters of reddenings derived using this equation with those from main sequence fitting from the literature. Thus, the probability of membership can be estimated from how closely the computed reddening of a given star agrees with the value accepted for the cluster. Examples of how this procedure can be applied are shown in Table I for two clusters. The columns of the table give the accepted reddening and distance modulus of the cluster, the observed magnitude of the star and the predicted spectral type if the star is a cluster member, and the parameters obtained from our DDO photometry. The values in parenthesis are those that result if the cluster reddening is used rather than the DDO determination. For NGC 2269 it is clear that star 1 is a foreground star and star 4 a probable cluster member. For NGC 2972 star 11 is probably a member, stars 2 and 14 background supergiants.

The next step is to identify which of the cluster members are the clump stars. No satisfactory solution to this problem has yet been found and so far the procedure followed has been to observe and reduce all stars in the neighbourhood of the clump and then make the selection on the basis of the agreement in the derived temperatures, gravities, and bolometric magnitudes.

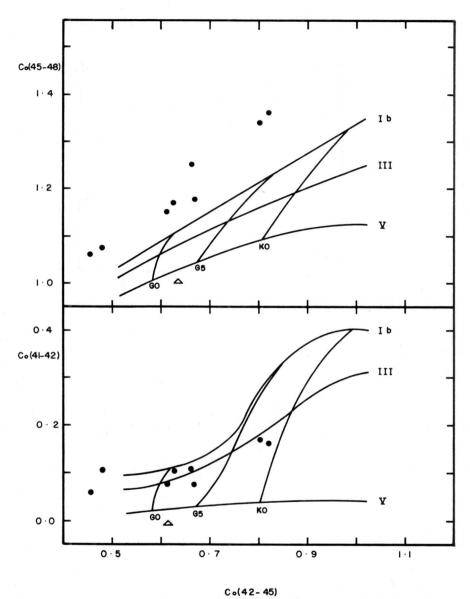

Fig. 1. Observed colours of several stars in M 92. One of the stars (triangle) is shown by the DDO colours to be a foreground G dwarf.

3. The Effective Temperature and Surface Gravities

Let us now assume that we can say with some certainty what the interstellar reddening of

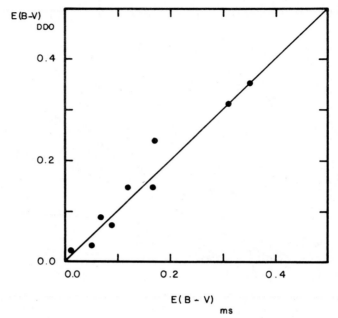

Fig. 2. A comparison between the reddening values found from DDO photometry with those from main sequence fitting for several open clusters.

TABLE I

NGC	$E(B-V)$	$V-M_v$	Star	V	Predicted sp. type	Results from DDO photometry			DDO Spectral type
						$E(B-V)$	$V-M_v$		
2269	0.19	10.7	1	8.6	G9 II	0.01	8.6	(7.2)	K1 III (K pec)
			4	8.2	G8 II	0.18	10.7	(10.7)	G5 Ib-II (G5 Ib-II)
2972	0.35	11.5	2	11.4	K1 II-III	0.30	17.4	(16.8)	G8 Ib (G8 Ib)
			11	12.0	G8 II-III	0.29	11.7	(11.1)	G8 III (G8 III)
			14	9.4	K3 II	0.45	15.1	(15.7)	K0 Ib (K3 Iab)

the cluster is and which are the physical members. We correct the observed DDO colours for reddening. The DDO $C(45-48) - C(42-45)$ two colour diagram has been empirically calibrated in terms of effective temperature and surface gravity (Figure 3) and therefore the unreddened indices give directly the desired physical parameters. The calibration, however, is only valid for stars of normal, i.e. approximately solar metal abundance. For stars with significantly non-solar abundance the colour indices must be corrected to account for the different degree of line blanketing:

$$COLOUR_{normal} = COLOUR_{observed} - A[Fe/H]$$

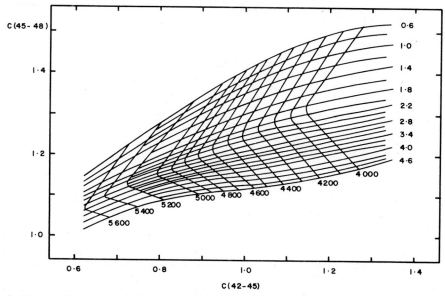

Fig. 3. The relationship between the DDO colour indices and effective temperature (T_{eff}) and surface gravity (log g).

The correction factors A have been empirically determined and tabulated by Osborn (1973). To determine [Fe/H] use is made of the DDO $C(41-42)$ index. Janes (1975) has tabulated the average value of this index as a function of $C(45-48)$ and $C(42-45)$ and has shown that the cyanogen anomaly δCN, defined as the difference between the observed $C(41-42)$ of a star and the normal value for the observed $C(45-48)$ and $C(42-45)$, is well correlated with [Fe/H]:

$$[Fe/H]_{DDO} = 4.5 \, \delta CN -0.2$$

This equation is used to derive [Fe/H] and the resulting value used to correct the observed $C(45-48)$ and $C(42-45)$ indices so that the temperature-gravity calibration can be applied. The derived [Fe/H] can also be adopted as an indication of the heavy element abundance Z.

4. The Helium Abundance

The effective temperature and surface gravity of the clump stars can now be used to obtain the helium abundances using Gross' theoretical results (Figure 4). It is seen in the figure that provided the mass of the star is less than about $1.0M_\odot$ the determination is insensitive to the core mass, the metal abundance, or the total mass. Evolution of the star is also parallel to the ZAHB lines. One can check if the mass of the observed star is less

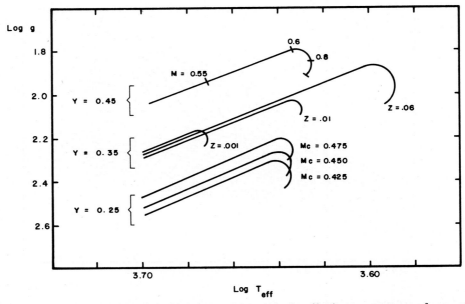

Fig. 4. The position of the zero age horizontal branch in the effective temperature-surface gravity plane based on Gross' models. The dependences on the helium abundance and on the total mass (upper), metal abundance (middle), and core mass (lower) are indicated. It is seen that the helium abundance is the dominant parameter.

than $1.0M_\odot$ if the distance modulus, hence the absolute magnitude of the star, is known. Combining the Stephan-Boltzmann equation and the definition of surface gravity an expression for the mass in terms of the effective temperature, surface gravity and absolute magnitude can be derived (Osborn, 1975).

$$\log M/M_\odot = \log g - 4 \log T_{eff} - 0.4 \, M_{bol} + 12.49$$

If the mass is in the permitted range then the helium abundance is determined by the adopted temperature and gravity.

A much stronger case can be made by requiring that the results be internally consistent. For example, not only must the derived mass be less than $1.0 \, M_\odot$ but it also must be greater than about $0.4 \, M_\odot$ or it is impossible to form a horizontal branch star. This consideration implies that once the surface gravity and effective temperature have been specified the bolometric magnitude and therefore the distance modulus of the cluster are fixed within rather narrow limits. If the data are consistent this distance modulus will (1) lead to a position of the star in the $M_{bol} - T_{eff}$ plane, which is also a function of Y as well as of Z, consistent with the previously adopted values and (2) also place the cluster main sequence in a reasonable position in the theoretical HR diagram.

The use of these restrictions to obtain the most consistent estimate of the cluster abundance can be illustrated by an example from a study of NGC 2420. DDO photometry has been published for 11 stars of the cluster of which five form part of the clump on the giant branch. The derived data for the five clump stars are listed in Table II. The accepted reddening, true distance modulus, and metal abundance of the cluster are (McClure *et al.*, 1974): $E(B-V) = 0.02$, $V_0-M_v = 11.4$ and $[Fe/H] = -0.4$ ($Z \approx 0.01$). The mean values for the five clump stars are

$$\log T_{eff} = 3.67 \qquad \log g = 2.4 \qquad M_{bol} = +0.7 \qquad M = 0.8 \, M_\odot$$

TABLE II

Star	Log T_{eff}	Log g	M_{bol}	$M/M_\odot$
1	3.671	2.0	+0.51	0.4
2	3.670	2.6	+0.69	1.3
3	3.670	2.7	+0.72	1.7
4	3.655	2.2	+0.73	0.6
5	3.672	2.4	+0.79	0.8

From the average temperature and gravity and Figure 4 we find a helium abundance of $Y \approx 0.25$. However, adopting $Z = 0.01$ we find from the position in the $M_{bol}-T_{eff}$ diagram (Figure 5) that $Y \approx 0.35$, in disagreement with the former value.

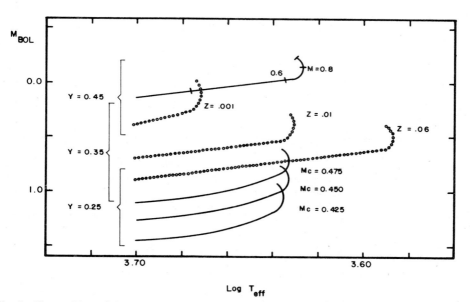

Fig. 5. The position of the zero age horizontal branch in the M_{bol}-effective temperature plane. As in the previous figure, the dependences on helium abundance and on total mass, metal abundance, and core mass are shown. In this diagram the effect of a change in helium abundance can be confused by changes in the other parameters.

Agreement between the two could be obtained by assuming a distance modulus about 0.4 smaller but this change would imply a helium abundance for the main sequence stars of $Y \approx 0.45$ (see Figure 6) and some of the derived masses become uncomfortably small. The same difficulties are encountered if one tries altering the adopted Z. A consistent case can be made by assuming that the two stars with high masses are red giants. Then the average values of the true clump, i.e. horizontal branch, stars are

$$\log T_{eff} = 3.67 \qquad \log g = 2.2 \qquad M_{bol} = +0.7 \qquad M = 0.6\,M_{\odot}$$

These values lead to $Y = 0.34$ (Figure 4) and $Y = 0.33$ (Figure 5) and the main sequence position is consistent with $Y = 0.35$ and $Z = 0.01$.

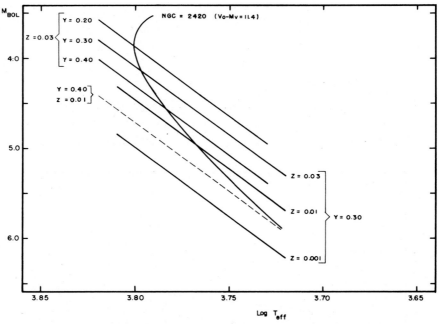

Fig. 6. The position of the zero age main sequence in the M_{bol}-effective temperature plane. There is a substantial dependence on both helium abundance and metal abundance. The three helium abundances are for $Z = 0.03$ while the three metal abundances are for $Y = 0.30$. Also shown is the observed sequence for NGC 2420 (McClure *et al.*, 1974) and, for comparison, the approximate theoretical position for a sequence with $Y = 0.40$, $Z = 0.01$ (dashed line).

5. Conclusions

In conclusion, we wish to repeat that the object of this paper has been to describe our work in general terms rather than present any definitive results. We conclude from our exploratory investigations that good (X, Y, Z) values for clusters can be obtained by looking for the interpretation of the observational data that is most consistent with the whole of the results of theoretical models. This result is not surprising since this has been exactly the approach used by several authors with respect to the Hyades. What is surprising is how infrequently these methods have been applied to other clusters.

In the present programme two clusters — M 67 and NGC 2420 — have so far been analyzed (Osborn, 1974, 1975). DDO photometry has been obtained for red stars in the fields of 13 additional clusters. The photometry shows that most of these stars are physical cluster members. These data will be discussed in a forthcoming paper.

References

Gross, P. G.: 1973, *Monthly Notices Roy. Astron. Soc.* **164**, 65.

Janes, K. A.: 1975, *Astrophys. J. Suppl.* **29**, 161.

McClure, R. D.: 1973, in C. Fehrenbach and B. Westerlund (eds.), *Spectral Classification and Multi-colour Photometry*, p. 162.

McClure, R. D., Forrester, W. T., and Gibson, J.: 1974, *Astrophys. J.* **189**, 409.

McClure, R. D. and Racine, R.: 1969, *Astron. J.* **74**, 1000.

McClure, R. D. and van den Bergh, S.: 1968, *Astron. J.* **73**, 313.

Newell, E. B., Rodgers, A. W., and Searle, L.: 1969a, *Astrophys. J.* **156**, 597.

Newell, E. B., Rodgers, A. W., and Searle, L.: 1969b, *Astrophys. J.* **158**, 699.

Osborn, W.: 1973, *Astrophys. J.* **186**, 725.

Osborn, W.: 1974, *Monthly Notices Roy. Astron. Soc.* **168**, 291.

Osborn, W.: 1975, *Monthly Notices Roy. Astron. Soc.* **172**, 631.

Philip, A. G. D.: 1972, *Astrophys. J. (Letters)* **171**, L51.

DISCUSSION

W. Bidelman: I did not quite understand the origin of the two discrepant values of Y that you quoted.

Osborn: The first value of Y comes from the mean position of the stars in the temperature-gravity diagram (Figure 4). The second value comes from the position in the temperature – M_{bol} diagram (Figure 5).

Andersen: I am impressed by the small range you find in Y. We have been working on an F-type eclipsing binary where we know the mass to about 2%, the radii to about 1% from very well-observed *uvby* light curves, the temperature from the accurate mean *uvby* indices, and the metal abundance from the m_1 index, and still we don't trust the resulting value of Y to be better than 0.05.

Osborn: Of course, I have not put any errors on my values. Let us say that the determination of $Y = 0.25$ from Figure 4 and $Y = 0.35$ from Figure 5 can be taken to mean that $Y = 0.30 \pm 0.05$. There seems to be no parameter that I can change that would give any Y value outside this range.

Williams: First, how many colours have you observed and how many parameters do you derive from them?

Secondly, have you observed any branch stars and derived abundances for them?

Osborn: We observe three colours C (45–48), C (42–45), and C (41–42). These are used to derive three physical parameters: effective temperature, surface gravity, and metal abundance. The helium abundance follows from comparison of these three with theoretical models.

No, I have so far observed only clump stars.

Bell: If you change the reddening, do the Y values become in better agreement?

Osborn: In this particular case of NGC 2420 the reddening is quite small, 0.01 or 0.02 depending on who you believe. Therefore, the reddening cannot be decreased significantly. Increasing the reddening increases the discrepancy.

PART IV

ABUNDANCE EFFECTS IN SPECTRAL CLASSIFICATION

ABUNDANCE EFFECTS IN SPECTRAL CLASSIFICATION

C. JASCHEK

Observatoire de Strasbourg, France

Abstract. The paper attempts to put together the spectroscopic definitions used in the literature for different groups of peculiar objects linked with abundance anomalies.
The groups discussed are:

> O-type subdwarfs
> CNO stars
> He-strong stars
> He-weak stars
> B-type subdwarfs
> Ap stars
> Am stars
> λ Boo stars
> δ Del stars
> Fm stars
> A-F-G stars with weak metal lines
> Stars with CN anomalies
> Stars with CH anomalies
> Ba stars

In each case the useful range in dispersion is quoted and bibliographic references are provided.

When a system of spectral classification is set up, one automatically divides the stars into those that fit into the established scheme and those that do not. The latter ones are the peculiar objects with regard to the reference frame used. (Outside of this frame the designation has no meaning, because nothing is peculiar by itself). This situation lasts until a new classification scheme redefines the reference frame, re-shifting also the peculiarities according to the new physical insights.

Such a situation can be illustrated nicely with a group of peculiar stars – peculiar with regard to the HD system – called the 'c' stars. With the advances in the physics it turned out that these were luminous stars, and the next classification scheme, that of Morgan *et al.* (1943) incorporated all the former 'c' stars into a consistent scheme. This situation illustrates very well the need for using both theory and classification to advance in the field. Classification can tell only what happens and can set apart stars for further study, whereas theory should be able to tell the 'whys'.

I must admit that this picture is an oversimplification, because it omits the contributions of the photometrists and the astrometrists. If often happens that a group of peculiar objects is first defined by one characteristic and named by it, but that later on other characteristics are found which bear on most of the stars – but not on all – of the group. For instance, it turns out to be quite a task to know what is really meant when calling a star Am. It could be so-called because of discrepancies between the spectral types

B. Hauck and P. C. Keenan (eds.), Abundance Effects in Classification, 113–125. *All Rights Reserved.*

assigned to the K-line, the Balmer lines and the metallic lines (i.e. because of its spectral classification); because of the ratio of two line-intensities (for instance Sc/Sr); because of the narrow band photometry performed, (for instance the Strömgren photometry); because of the multicolour photometry (*UBVRI*) or because of an abundance analysis. A star called Am by one criterion is not necessarily so-called according to the others.

This is a very unfortunate situation because it gives a wrong point of departure for many investigations. In the particular case quoted, it is very difficult to single out Am stars photometrically, if there is no clear spectroscopic definition of what an Am star is.

Because of this situation, I will use my time to give a summarized version of the spectroscopic definitions of the various groups which are mentioned when talking of abundance efects. I have thought that such a documentation might be useful especially for the non-spectroscopists.

I said specifically 'spectroscopic definitions', because I think one can get a very clean description – and definition – just from the spectrum. I will thus try to avoid definitions implying theory; for instance I will not mention gravity or temperature. This is not only because I am not a theoretician, but because I think it is best to describe spectra such as they are observed. Take for instance luminosities of stars. Today everybody is used to roman numeral luminosity classes, without reference to the absolute magnitude related to that class. By taking 'luminosity' out of the hands of those working with distance scales, I think we have succeeded in concentrating more on what is seen in the spectrum. Because of the purely spectroscopic definitions, I will also exclude group assignments based on velocity, be it radial or space velocity. I wil therefore also not enter into the various population type assignments, except if they can be established from only an examination of the spectrum.

Therefore, in a certain sense, I am shrinking away from 'understanding' observations. This is true for all morphological science, be it botany, zoology, crystallography or spectral classification, and I accept voluntarily this self-imposed limitation. I think I can still speak about Am stars, despite ignoring if the effects are caused by abnormal abundances due to nuclear reactions, or to an abnormal atmospheric structure, or to diffusion processes, or to accretion, or to mass exchange between binary star components.

I mention this explicitly because since I am speaking about 'abundance effects in classification' I should know (but I don't) if all the groupss I will be mentioning are really due to anomalous abundances. I think that in many cases this is likely, but we are not always sure, and it is for this reason that I shall include groups which perhaps some colleagues would not include if they had to talk about the subject.

I can be more definite about what I shall not include, for instance variable stars like Cepheids, RR Lyrae and novae. I will also omit binaries in which there are interactions between the components. I shall omit also stars which belong to the main groups of classification, like white dwarfs, carbon and S stars and WR objects.

A question I would like to discuss before going into details is that of the line strengths. You will see that in many groups one speaks of a line – or lines – as being weaker than in a standard star (or in a 'normal star'). The degree of deviation from normalcy can be defined, like in the MK system, by means of a group of standard stars. This procedure has usually the difficulty that the list of standards is small, that the standards are not observable from the part of the sky and at the time of the year you are observing and that they do not cover the anomaly at all spectral types and luminosities. Another procedure is to refer the anomalies to another type of stars. One can say for instance that the spectrum is that of an A5 giant, but that the line at $\lambda 4215$ is enhanced as in an F5 III star. This procedure has the advantage that one can omit the selection of new standards for the Sr II line strength. Obviously the procedure fails if one tries to apply it to an exotic object like a mercury star, because the Hg line exists only in these stars, and not in the standards.

The last question I would like to mention is that each group is defined by what is seen at a given dispersion, so that it happens that at some dispersions the peculiarity becomes blurred and later on disappears when one uses different dispersions. I will thus quote what happens at each dispersion. The dispersion is given in Å mm^{-1} at $H\gamma$. I will add that there are some reasonable doubts if dispersion alone tells the whole story — if so, slit spectra and objective prism spectra of the same dispersion should give the same amount of information. However, this is only true if the spectra are also well enlarged, well exposed and if the seeing was good at the time the objective prism spectra were taken. I will not, however, go into more detail and assume simply that the dispersion by itself is the most important parameter.

Since Dr Seitter will describe in her paper what can be made out at dispersions between 240 and 1300 Å mm^{-1}, I will restrict my description mostly to what can be seen at dispersion from 40 to 200 Å mm^{-1}.

Finally I would like to point out that the bibliography indicated in each group is definitely incomplete. I have simply chosen those papers I found best suited to my purpose and I ask you to excuse any omission.

And now I will start with the descriptions.

1. O-Type Subdwarfs (sdO)

Balmer lines are broad when compared with dwarfs of similar spectral type; the number of Balmer lines is also smaller ($n\sim10-12$, compared to $n\sim18$ in dwarfs). Furthermore, N IV ($\lambda 3479$ for instance) is abnormally enhanced. If the ultraviolet spectrum is visible, group members can be detected at 180 Å mm^{-1} (Sargent and Searle, 1968). Greenstein and Eggen (1966) called attention to a possible division into line-rich and line-poor stars. Attention should be paid to the ratio $\lambda 4200 : \lambda 4686$ of He II (Sargent and Searle, 1968) which might be smaller than in dwarfs. But on the whole He is normal or even stronger than in dwarfs (see also Graham and Slettebak, 1973). Another important element to examine, is S III.

These stars can be distinguished from B-type subdwarfs by the presence of $\lambda 4686$

He II. This line is generally not present in B-type subdwarfs. See, however, Baschek (1975) for two exceptions.

Several dozen of these stars are known.

2. CNO Stars

Under this name, we will group early type stars (O-B5) in whose spectra the lines of either C, N and/or O are of abnormal intensity. The first stars of this group were described by Jaschek and Jaschek (1967); the group as such was defined by Walborn (1970) at 60 Å mm^{-1}. Although some stars can be detected even at 120 Å mm^{-1}, higher dispersions should be used.

Since C, N and O do have a positive luminosity effect (i.e. enhance in supergiants) CNO rich stars are easier to detect in the dwarfs and CNO deficient stars in supergiants.

It is possible that the group could be subdivided into 'carbon' and 'nitrogen' subgroups (Walborn, 1971).

A few dozen stars of this type are known.

3. He Strong Stars

(Also called simply 'He stars'). They are by definition stars in which He is very enhanced with respect to hydrogen, which is either weak or absent.

If the hydrogen is weak, but still exists, the He I spectrum shows singlet / triplet ratios similar to those in B-type dwarfs (for instance λ 4387/λ 4471).

If the hydrogen is completely absent, no Balmer jump exists and the singlet / triplet ratios are different from those in dwarfs. Oxygen lines are usually absent in these objects and C is enhanced. This group is also called 'H-poor'.

The He-group was apparently first defined by Bidelman (1952). Further studies at high dispersion (by Hunger and co-workers, see Hunger 1975) have added many details, but no other outstanding common characteristics of the group.

About thirty stars of this group are known.

4. He Weak Stars

They can be defined as B-type stars in which the hydrogen-line spectrum and the helium spectrum disagree, the latter being weaker than in dwarfs of similar hydrogen-line spectrum. This definition is incomplete since many early Ap stars also do have weak helium lines. One should therefore add the condition '... and no strong Si II lines are visible'. The group was defined by Garrison (1967) on 90 Å mm^{-1} plates; Sargent and

Searle (1968) at 50 Å mm^{-1} define a group of Bw stars, whose definition is practically equivalent to the present one, except that their definition is not purely spectroscopic. At the lower end one can distinguish stars of this type at 180 Å mm^{-1}.

A subdivision of the group has been attempted by Jaschek and Jaschek (1974) and Baschek (1975) into:

(1) αScl type stars. Ti II and Fe II are visible together with a hydrogen line spectrum corresponding to about B5.

(2) P-stars. Stars in which P II lines are very enhanced.

(3) Blue helium stars. Si II is slightly stronger than normal.

Dispersions of at least 40 Å mm^{-1} are needed to distinguish members of these groups.

Attention is called to the possibility that many of the stars of this group are spectrum variables.

One important point here is that stars of this group can belong either to Population I or to Population II. In the latter case they are called horizontal branch stars. Greenstein and Sargent (1974) referring to the latter, state that 'either one or more of He I, C II and Mg II are weakened; generally He I is weak', and that the rotational velocity is low. Sargent and Searle (1968) stress that He I weakness does not correlate with the behaviour or the lines of heavier elements.

At the present is does not seem possible to distinguish in a purely spectroscopic way whether a star belongs to the Population I or II. In particular the Balmer line spectrum does not give any clue. Incidentally this group provides a good example of the interaction between definitions and theory. Horizontal branch stars in the solar neighbourhood are picked out as such because of their He-weakness. But it is argued sometimes that all horizontal branch stars are helium weak, which might be true but cannot be ascertained from material selected in the way described.

About 10^2 objects of this type are known.

5. B-Type Subdwarfs (sdB)

In this group, Balmer lines are too broad for the spectral type and the number of Balmer lines is smaller than in dwarfs ($n \sim 10-12$, as compared with $n \sim 18$ for normal dwarfs). Often, but not always, the He I lines are weak.

The group was defined by Greenstein and Münch around 1954, but I have been unable to trace the exact origin. Members of this group can be detected on 180 Å mm^{-1} plates (or lower), if the Balmer jump is visible. When observed at 50 Å mm^{-1}, Sargent and Searle (1968) call attention to anomalies in the singlet/triplet ratio of the helium lines. So, for instance, λ 4387/λ 4471 is smaller than in dwarfs.

Greenstein and Sargent (1974) studied a sample of these stars at higher dispersion (around 20 Å mm^{-1}). Several anomalous objects were found, but no regularity emerges. Perhaps the λ 4253 S III is important for drawing subgroups.

About 10^2 objects are known.

6. Ap Stars

They were described as a group for the first time by Morgan (1933) (using $30\,\text{Å}\,\text{mm}^{-1}$ spectrograms) as being late B- or early A-type stars in whose spectrum some lines are very enhanced. On the basis of the elements whose lines are most intensified, Morgan introduced six main groups, namely the Mn, λ 4200, Cr, Eu, Sr and Si stars. This sequence is given in order of advancing spectral type, except for the Si-stars, which are scattered through the whole group. In recent years one more group, that of the Hg stars, has been introduced (Bidelman) and the sequence has been slightly rearranged. Also recent authors tend to list all elements whose lines are enhanced, not only the most enhanced one.

The following groups can be regarded as being the classical ones. They are listed in order of advancing spectral type and some of the most characteristic lines of each element are provided.

Si – λ 4200	λ 4200, 3955
Mn (II)	λ 4137, 4206 also λλ 3441, 3460, 3474
Hg (II)	λ 3984
Si (II)	λ 4128–31, 3854, 56–62
Cr (II)	λ 4171, blend around 4111, 4233
Eu (II)	λ 4129, 4205, 3930
Sr (II)	λ 4077, 4215

Not all of these families can be detected efficiently at all dispersions. Although extreme cases of Si, Cr and Sr stars can be detected at $300\,\text{Å}\,\text{mm}^{-1}$ (Honeycutt and McCuskey, 1966), the groups become reasonably well defined only at $120\,\text{Å}\,\text{mm}^{-1}$ (Cowley et al., 1969) except for the Mn and Hg stars which are still difficult. Dispersions of at least $60\,\text{Å}\,\text{mm}^{-1}$ (Osawa, 1965) should be used to overcome this. The Mn-group is easier to detect in the ultraviolet region, through the lines of M.3 (λ 3441, 3460, 3474) (Nariai, 1967).

In the last decades the simple classification scheme outlined has become more complicated because at higher dispersion more intermediate groups were added. So Osawa (1965) uses 19 groups, most of them being combinations of the seven listed above.

With regard to other characteristics of the Ap stars, the hydrogen-line spectrum seems to be comparable to that of dwarfs of similar type, both in the line shape and the Balmer jump. If helium lines are visible, they tend to be weak for the type assigned to the hydrogen lines (Deutsch, 1956).

Attention must be called to the large percentage of spectrum variables in this group, which makes impossible the assignment of a unique classification of each object, unless non-variability is proven.

Attempts have been made to simplify the subdivisions within this group (see for instance Jaschek and Jaschek, 1974). Usually the subdivision is made into Si stars, Mn stars and Eu-Cr-Sr stars (also called rare earth objects).

It should be added that it is in the latter group where most of the outstanding composition anomalies are found, such as the presence of very heavy elements, up to uranium (Jaschek and Malaroda, 1970), and the elements characteristic of nuclear fission (see review by Kuchowicz, 1973).

The number of known Ap stars is large ($n > 1 \times 10^3$).

7. Am Stars

They were discovered by Titus and Morgan (1940) in the Hyades and several of them were described in the Atlas (Morgan *et al.*, (1943). At 100 Å mm^{-1} one can classify them according to the Ca II K line, to the hydrogen lines and to the metallic lines. If one calls these types respectively $S(K)$, $S(H)$ and $S(m)$, one has $S(K) \leqslant S(H) \leqslant S(m)$.

Furthermore the metallic lines are sharp and some lines usually strengthened in giants (λ 4077 Sr II, Fe II λ 4173 = λ 4178) are also enhanced in the Am's. This pseudo--luminosity effect resulted in these objects being classified as giants. However there are differences from the spectrum of true giants, especially λ 4417 $\ll \lambda$ 4481. (Cowley *et al.*, 1969).

This description applies to dispersions between 100 and 40 Å mm^{-1} but at higher dispersions care must be taken to compare with stars of low rotational velocity.

Stars with marginal Am characteristics are often called 'mild Am'.

At 20 Å mm^{-1} higher, the ratio λ 4246 Sc II/λ 4215 Sr II has been used (Bidelman, 1956; Conti, 1965) to characterize Am stars. It must, however, be stressed that not all Sc II weak stars are Am stars (in many Ap stars of the Cr-Eu-Sr type, Sc II weak) and that not all Am stars are Sc II weak.

At lower dispersions, one can pick out extreme Am stars (with $S(K) \ll S(m)$) up to 300 Å mm^{-1} (Honeycutt and McCuskey, 1966).

Many Am stars are known ($n > 10^3$).

8. λ Boo

The star λ Boo, prototype of the group, was first described by Morgan *et al.* (1943) in the Atlas, at 110 Å mm^{-1}. For the Balmer line strengths, all other metallic lines are abnormally weak, including those of Ca II.

At higher dispersions it becomes evident that the lines are not sharp and that the velocity is small; Slettebak *et al.* (1968) used this as a discriminant from horizontal branch stars. For a description of the group at higher dispersion see Baschek and Searle (1969).

Less than ten objects are assigned to this group.

9. δ Sct

The first object of this group was described by Roman (1951), but most of the stars were discovered photometrically later on. Their spectra seem about normal for early F-type giants except for the weakness of the Ca II lines. (Bidelman, 1951). Morgan and Abt (1972), who have studied spectroscopically for the first time a large sample of δ Sct Stars, state that these stars have '.. Luminosity classes brighter than class V and that the spectra of these δ Sct stars do not form a homogeneous group. (The Ca II lines show a great range in intensity among objects having similar spectral types and luminosity classes)'. These stars can be detected between 40 Å mm^{-1} and 120 Å mm^{-1}.

A few dozen stars of this type are known.

10. Fm Stars

Bidelman discovered the first star of this type, studied by Preston (1961). Later on a few more stars were discovered at 80–100 Å mm^{-1}, characterized by a metallic-line spectrum which corresponds to a later type than that indicated by the hydrogen lines. The G band corresponds to an even more advanced type than the metallic lines. Generally Sr II is strong and Ca II slightly earlier than the hydrogen lines.

The denomination Fm introduced by Houk (1975) points to the relation with Am stars. There are reasonable doubts if this is really an independent group.

A few dozen stars of this type are known.

11. A, F and G Dwarfs with Weak Metal Lines

Morgan *et al.* (1943) mentioned for the first time stars in which the metallic lines are weaker than normal. Later on, Roman (1954) described the existence of a group of weak-lined F stars. She added that the behaviour of the hydrogen spectrum, including the number of visible lines and the jump, were all compatible. She studied then a sample of bright F5-G5 stars in which she found the so-called 'weak-line stars' defined at 125 Å mm^{-1} (Roman, 1952) as stars having weak metals, but no specific definition was provided. Morgan (1958) called attention to the fact that the procedure is extremely sensitive to misclassification. If for instance an F7 star is called F8, it might be called later on 'metal weak'.

Greenstein (1960) states
the characteristic feature of the spectra of these stars is the extreme weakness of metallic lines for the temperature given by the star's colour and by the level of excitation and ionization. The hydrogen lines are strong and have relatively sharp and deep cores. Turbulence must be small from the few data given so far by curves of growth and from the visual appearance of extreme line sharpness. Rotation is small or absent.

Although this is not a purely spectroscopic description, it indicates that the essential feature is the discrepancy between the spectral type corresponding to metallic lines and to the hydrogen lines.

It should be added incidentally that some, but not all, RR Lyrae stars are metal weak.

Bond (1970) made a survey of metal deficient, i.e. weak metallic line, stars. He selected them from a sample of stars found on objective prism plates, at 110 Å mm^{-1}. He found that the easiest objects to find are those of type F and G; at spectral type A excellent seeing is needed, to prevent the faint lines from disappearing. He analyzed then some objects on slit spectrograms, with dispersions ranging between 77 and 142 Å mm^{-1}. The criteria he used are

... to classify the stars primarily on the basis of their hydrogen line strengths. In some cases the strength and appearance of the G band were also taken into account.... Luminosity classes were assigned primarily on the basis of the ratio of Sr II λ 4077 to the nearby Fe I lines and, in the earlier stars, by the ratio of the blend near λ 4172 (mostly Fe II) to Ca I λ 4226. ...Once a star had been classified, its spectrum was again compared with the standard of the same type (if available) and the degree of weakening of the metallic lines (principally in the range λλ 3933–4226) was estimated. Four adjectives were used – none, slight, moderate, extreme.

With lower dispersions, extreme cases can still be found at 200 Å mm^{-1} according to Graham and Slettebak (1973), like the K2 star mentioned in the publication.

On the other hand, Wilson (1961) used a dispersion of about 10 Å mm^{-1} to analyse possible difference in late G- and early K-type dwarfs. He found two characteristics – namely that some stars do possess emissions in the core of the Ca II lines and second, that for equal intensity of the hydrogen lines, the strength of the metallic lines can be very different. Stars exhibiting these phenomena are probably related to the more extreme cases found with lower dispersion at earlier spectral types.

Several hundreds of stars of this type are known.

12. CN Anomalies

Although CN had been used as a luminosity criterion by Lindblad back in 1922, it was Morgan and Keenan who pointed out in the Atlas (1943) that both CN and metals were weak in some high-velocity stars. The CN effects are noticeable in giants of type G7-K2, because it is in this region where the CN bands are normally strong enough to become conspicuous features of the spectrum, whereas CN is never that conspicuous in dwarfs.

The CN strong stars were discovered by Morgan and Nassau on objective prism plates; they were later called λ 4150 stars by Roman (1952).

Since usually the λ 4216 CN band is more easily visible than both the λ 4150 and/or the λ 3889, Keenan and Keller (1953) used at 100 Å mm^{-1} the ratio λ 4216/ λ 4172. λ 4216 is a blend of λ 4215 Sr II and λ 4215 CN, whereas λ 4172 is a blend of metallic lines. When λ 4216 is weak (i.e. CN weak), then the index is small. Keenan (1961 and 1964) recommended then the introduction of the 'CN discrepancy' = $CN_* - \overline{CN}$ which is sometimes also called the 'CN anomaly'. He thinks that this index is a consistent indicator of metal abundance in stars later than G5. (Morgan and Keenan, 1973). A list of stars with various degrees of CN discrepancy (between +3 and −3) defines the observational status. One should observe that Keenan defines the strength of λ 4215 on the basis of the intensity ratios:

I (4211−13) : I(4219−21), whereas Schmitt (see below) uses
I (4210−16) : I(4215−27)

What photometrists measure when dealing with CN is something different and it is well illustrated in Golay (1975).

Schmitt (1971) used the following definition of a 'CN strong' star − '... is a star whose spectrum shows more absorption in the 0−1 seq. of the CN band with its head at λ 4216 than the average MK standard shows at the same spectral type and luminosity class'. His stars were selected from a list of λ 4150 stars found on objective prism plates at 120 Å mm^{-1} and observed later on slit spectrograms (∼ 100 Å mm^{-1}).

Later on, CN strong stars were called metal rich and CN weak stars either population II stars, or metal poor stars. While this might be right, it is also true that it is best to adhere strictly to what is observed (CN anomalies) rather than to generalize. This is also important in view of the fact that there exists at least one instance in which a dwarfs has weak metallic lines and strong CN. (Wilson, 1961, at 10 Å mm^{-1}).

It should be added finally that CN strong stars tend to have strong Ca I λ 4226, a strong G band and sometimes weak Hγ.

With regard to dispersions, a range of 80−150 Å mm^{-1} can be used to detect these stars. A large number of objects is known ($n > 10^3$ stars).

13. CH Anomalies

We will group here all stars in which the CH bands have an anomalous intensity. Usually it is the G band which is most conspicuous and usually these stars are giants, of the types G-K.

The first object of this group was found by Keenan and Morgan (1941) at 120 Å mm^{-1}; the group as such was described by Keenan (1942) who called them CH stars. These stars are characterized by very strong bands due to CH, including an extremely strong G band, and considerably weaker lines of neutral metals than in typical 'carbon' stars. The features are so strong as to suppress even the λ 4226 Ca I line. Usually, but not always, C_2 and CN are strong in these objects. λ 4077 and 4215 of Sr II are always strong and usually λ 4554 BaII is present. These stars are closely related to the carbon stars.

Later on N. Houk (1975) found at 110 Å mm^{-1} two or three dozens of such stars, although the enhancement is less intense than in the previous group.

On the other hand there exists also a group of stars in which CH is very weak or absent. This is accompanied usually by very weak hydrogen lines and strong C_2 and CN bands and strong CI lines. The class was first described by Bidelman (1953) at 80 Å mm^{-1}. These stars are also called 'hydrogen-poor carbon stars'.

Recently Bond (1974) added a new subgroup, namely that of the subgiant CH stars. He describes them as G type spectra with weak metallic lines but enhanced features of CH and strong Sr II λ 4077 and 4215 lines. In some extreme cases, the metallic lines are so weak that the Sr II lines, the CH features, the Balmer lines and sometimes λ 4554 are the only features clearly seen longward of the Ca II K and H lines.

Stars of this type are also detectable at 110 Å mm^{-1} (Houk, 1975). Less than a hundred 'CH anomalous' objects are known.

14. Ba Stars

A K-type giant with a strong λ 4554 Ba II line is called a barium star. The group was first described by Bidelman and Keenan (1951) at 80 Å mm^{-1} although isolated examples were known before. These authors add that all the Ba II stars show, in addition '... a definite enhancement of the G band and also, probably, of the violet bands of CN as well. The lines of Sr II at λ 4077 and λ 4215 (the latter blended with the head of the (0.1) CN band) are also strengthened'. Also the C_2 band λ 5165 is stronger than in normal stars.

At higher dispersion most of the elements heavier than strontium are overabundant. Warner (1965) at 80 Å/mm introduced a notation of Ba II enhancement, with steps from 1 to 5, the latter one being the most extreme enhancement.

McConnell *et al.* (1972) at a dispersion of 110 Å mm^{-1} used the criterion of the 'prominence of the Ba II line at λ 4554, the strength of λ 4077 relative to Hδ, and the strengths of the CN band (λ 4215) and the G band.'

They use no 'Ba strength' notation, but use the terms 'certain' and 'marginal' Ba II stars.

They also call attention to a few stars which are probably borderline cases between C and Ba II stars. Another subgroup might be one of 'weak lines in their Fe group elements, when compared with MK standards and other Ba stars'.

Several hundreds of objects are known.

Acknowledgements

Finally I would like to thank my colleagues W. P. Bidelman, N. Houk and Mercedes Jaschek for their friendly criticism during the preparation of this paper.

References

Baschek, B.: 1975, in *Problems in Stellar Atmospheres and Envelopes,* Springer-Verlag.
Baschek, B. and Searle, W. L. W.: 1969, *Astrophys. J.* **155**, 537.

Bidelman, W. P. and Keenan, P. C.: 1951, *Astrophys. J.* **114**, 473.

Bidelman, W. P.: 1951, *Astrophys. J.* **113**, 304.

Bidelman, W. P.: 1952, *Astrophys. J.* **116**, 227.

Bidelman, W. P.: 1953, *Astrophys. J.* **117**, 25.

Bidelman, W. P.: 1956, private communication.

Bond, H. E.: 1970, *Astrophys. J. Suppl.* **22**, 117.

Bond, H. E.: 1974, *Astrophys. J.* **194**, 95.

Conti, P. S.: 1965, *Astrophys. J.* **142**, 1594.

Cowley, A., Cowley, C., Jaschek, M. and Jaschek, C.: 1969, *Astron. J.* **74**, 375.

Deutsch, A.: 1956, *Publ. Astron. Soc. Pacific* **68**, 92.

Garrison, R. F.: 1967, *Astrophys. J.* **147**, 1003.

Golay, M.: 1974, in *Introduction to Astronomical Photometry* , D. Reidel Publ. Co., p. 254.

Graham, J. A. and Slettebak, A.: 1973, *Astron J.* **78**, 295.

Greenstein, J. L.: 1960, in *Stellar Atmospheres*, Univ. of Chicago Press, p. 707.

Greenstein, J. L. and Eggen, O.: 1966, *Vistas Astron.* **8**, 63.

Greenstein, J. L. and Sargent, A. I.: 1974, *Astrophys. J. Suppl.* **28**, 157.

Houk, N.: 1976, this volume, p. 127.

Hunger, K.: 1975, in *Problems in Stellar Atmospheres and Envelopes*, Springer-Verlag.

Honeycutt, R. K. and McCuskey, S. W.: 1966, *Publ. Astron. Soc. Pacific* **78**, 289.

Jaschek, M. and Jaschek, C.: 1967, *Astrophys. J.* **150**, 355.

Jaschek, M. and Malaroda, S.: 1970, *Nature* **225**, 246.

Jaschek, M. and Jaschek, C.: 1974, *Vistas Astron.* **16**, 131.

Keenan, P. C.: 1971, *I.A.U. Trans* **9**, 404.

Keenan, P. C.: 1964, *I.A.U. Trans* **10**, 447.

Keenan, P. C. and Morgan, W. W.: 1941, *Astrophys. J.* **94**, 501.

Kuchowicz, B.: 1973, *Quart. J. Roy. Astron. Soc.* **14**, 121.

MacConnell, D. J., Frye, R. L., and Upgren, A. R.: 1972, *Astron. J.* **77**, 384.

Morgan, W. W.: 1933, *Astrophys. J.* **77**, 330.

Morgan, W. W., Keenan, P. C., and Kellman, E.: 1943, *An Atlas of Stellar Spectra,* Univ. of Chicago Press.

Morgan, W. W.: 1958, in D. J. K. O'Connell S. J. (eds), *Stellar Populations,* Vatican Observatory, p. 263.

Morgan, W. W. and Abt, H. A.: 1972, *Astron. J.* **77**, 35.

Morgan, W. W. and Keenan, P. C.: 1973, *Ann. Rev. Astron. Astrophys.* **11**, 29, Palo Alto.

Nariai, K.: 1967, *Publ. Astron. Soc. Japan* **19**, 180.

Osawa, K.: 1965, *Ann. Tokyo Astron. Obs.* (2) **9**, 123.

Preston, G. W.: 1961, *Astrophys. J.* **134**, 797.

Roman, N.: 1951, *Astrophys. J.* **113**, 705.

Roman, N.: 1952, *Astrophys. J.* **116**, 122.

Roman, N.: 1954, *Astron. J.* **59**, 307.

Sargent, W. L. W. and Searle, L.: 1968, *Astrophys. J.* **152**, 443.

Schmitt, J. L.: 1971, *Astrophys. J.* **163**, 75.

Slettebak, A., Wright, R. R., and Graham, J. A.: 1968, *Astron. J.* **73**, 152.

Titus, J. and Morgan, W. W.: 1940, *Astrophys. J.* **92**, 257.

Walborn, N.: 1970, *Astrophys. J.* **161**, L149.

Walborn, N.: 1971, *Astrophys. J.* **164**, L67.

Warner, B.: 1965, *Monthly Notices Roy. Astron. Soc.* **129**, 263.

Wilson, O. C.: 1961, *Astrophys. J.* **133**, 457.

DISCUSSION

Bidelman: Bond's supposed luminosity class IV CH stars do indeed appear to be nuclearly-evolved stars of quite low luminosity. They can also be considered hotter and less luminous Ba II stars. They raise very important questions.

With regard to the so-called rare-earth Ap stars, it should be emphasized that there are appreciable differences between the species of rare earths that are overabundant in the Sr-Eu stars like γ Equulei and the Cr-Eu stars like β Coronae Borealis. This difference should not be forgotten.

The Ba II stars published by MacConnell and Upgren, which I had some hand in, probably include quite a few stars similar to Bond's low-luminosity CH stars. Eggen has done photometry of some of the new Ba II stars and has concluded from this that many may not really be Ba stars. This conclusion is probably not entirely warranted in view of the above.

Williams: With reference to Dr Bidelman's comment on Eggen's photometry of the Ba II stars, he may like to know that I have carried out a spectroscopic analysis of one Ba II star not supported as such by Eggen's photometry and confirmed its overabundance of barium and other heavy metals. I do not think broad band photometry is a good discriminant of Ba II stars.

Walborn: I would like to comment on the complexity of the phenomena within the He-strong group. First, among the 'intermediate' group (which I prefer to call 'helium-rich' to avoid an implication of continuity with the extreme group), there appears to be a continuous range of helium enhancement, from slightly stronger than in normal stars, to almost as strong as the hydrogen lines. Secondly, within the extreme group there is a large range of gravities and also of CNO-line behaviour. The gravities appear to range from intermediate between those of subdwarfs and normal main-sequence stars, to as low as those of giants. While I think most extreme helium stars have C-enhanced, there are also examples of both C and N strong (HD 160641) and of N enhanced (BD + 13° 3224). So there are probably objects with a variety of evolutionary histories in this category.

Hack: Could you say what is the rotational velocity of He-weak stars on the average?

Jaschek: Low. Usually below 100 km s^{-1}.

Cayrel de Strobel: You said that you do not link population criteria, but if you put a 'sd' or VI on an F star, you do.

Walborn: The problem with luminosity class VI is that it has not been systematically defined. However, if it is used, it should mean that there is some luminosity criterion *in the spectrum* which indicates a luminosity lower than class V.

Morgan: (to Dr. Cayrel): Can you distinguish subdwarfs from dwarfs exclusively from spectroscopic means?

Foy: First, we can obtain the effective temperature using T_{eff} of a given star by hydrogen line profiles Hα or Hβ (if the star is not too cool).

Second, we obtained a relation between T_{eff} and the gravity g by imposing ionization equilibrium, i.e. we determine T_{eff} and $\log g$ so that for a given element, the abundance derived from neutral lines agrees with that observed from ionized lines.

Third, we obtained another independent relation between T_{eff} and $\log g$: it comes by imposing to find the same abundance for a given element either from weak lines of the curve of growth or from wing profiles of strong lines of this element. Magnesium is a good candidate to this criterion, because wings of Mg Ib triplet are very sensitive to gravity.

The intersection of these three criteria gives us T_{eff}, $\log g$ and [M/H]: the value of $\log g$ indicates to us whether the star is a subdwarf ($\log g > 4.80$ at solar spectral type) or not.

Baschek: I would like to comment on a problem concerning the *spectroscopic* distinction of sdO from sdB stars. The strength of He II 4686 depends on the temperature as well as on the helium abundance. There are some stars ('OB-type subdwarfs' HD 149382, Feige 66) which exhibit weak (neutral) He which is characteristic of sdB, but also He II 4686 which is characteristic of sdO.

Jaschek: Thanks for your observation. This is another case of objects intermediate between two groups.

HD STARS SOUTH OF δ = −53° HAVING PECULIAR ABUNDANCES: STATISTICS AND NOTATION

N. HOUK and M. R. HARTOOG

Department of Astronomy, University of Michigan, U.S.A.

Abstract. The HD stars between $\delta = -90°$ and $-53°$ have been classified on the MK system, using Michigan 4°+6° objective-prism Schmidt plates (108 Å mm⁻¹) taken at Cerro Tololo, Chile, and the results are published in Vol. 1 of the *University of Michigan Catalogue, Catalogue of Two-Dimensional Types for the HD Stars*. First, various statistics on the 32 500 normal stars and 2500 peculiar stars found are presented. Peculiar/normal ratios for various types of stars are given, along with discussion of the assumptions made. About 40% of the peculiar stars are of the Am or Ap types, and these will be covered in detail. Secondly, what notation to use for spectral types of stars with peculiar abundances is discussed and an explanation given of the notation used in Vol. 1 of the *Michigan Catalogue* for Am, Ap, strong and weak CN, weak metal, and other categories of peculiar stars.

DISCUSSION

Osborn: What about the 4% of the stars that could not be classified? Was this due to identification problems or to overlap problems?

Houk: These stars were almost all either too badly overlapped or too underexposed. They are still listed in the catalogue, with previously published type, usually HD types, given. A few stars were too overexposed to classify on our plates and Garrison and Hagen kindly provided spectral types for stars brighter than 4.75. We have had no insoluble identification problems so far; we used transparent computer plots of the HD stars to Schmidt scale, which are placed under the plate during classification.

Hack: I was surprised to see such a low percentage of Ap stars. Did you include He-weak, Mn stars, hot Si (λ 4200) stars among the Ap stars?

Houk: Only the Si (λ 4200) stars. The others either weren't looked for or don't have a spectral type containing 'Ap' which was the code the computer searched for in forming this list.

Rudkjøbing: When in the Am case we find the ratio by numbers of these 'peculiar' stars to the normal ones to be about 30 to 70, then an inclusion of very mild cases might perhaps change the ratio to that of about equally numerous groups. It would then be questionable which of these two groups should in fact be termed 'peculiar'.

Houk: The ratio is 30 to 100. It is true that within the A5-A9 IV-V range the metallic-line characteristic is very widespread.

Keenan: (1) These statistics are very valuable because they provide reliable percentages for many of the groups for the first time.

(2) The small number of barium stars suggests the question: How weak can the Ba line be on Warner's scale and still be detected on your plates?

Houk: I don't know what they would be on Warner's scale. Other characteristics such as strong Sr, CN and CH in the barium stars are more likely to be noticed than the Ba line, though I found a couple of late Ba stars that did have the 4554 line without showing these other characteristics.

Morgan: I imagine that Hg Mn stars would be very difficult to recognise on your plates.

Houk: Yes, they are. At first I tried to do so, so several such types will be found in the catalogue, but I soon stopped even trying to pick them out. Therefore no statistics should be done on these.

Mendoza: How many stars show Ca II in emission?

Houk: I don't know. In this table all the '*e*' stars are counted but most Ca II emission stars don't

B. Hauck and P. C. Keenan (eds.), Abundance Effects in Classification, 127–128. All Rights Reserved.

have H in emission so tradition is followed and a 'p' added to the spectral type rather than an 'e' for Ca II emission.

Bidelman: When the desirability of reclassifying the HD stars was first brought up at the Saltsjöbaden conference, a well-known astronomer commented that this was all very well but that he doubted that any man could be found to do the job. Well, we found a woman!

Houziaux: Do you have an idea of the minimum equivalent width you can detect with your objective prism spectra? This may affect a great deal your statistics on peculiar objects. Furthermore, for emission-line objects, as the Hα/Hβ intensity ratio is usually fairly high, there may be many more emission-line objects than the ones where you can detect Hβ emission.

Houk: I really have no idea. For a number of the stars for which I remarked that Hβ was possibly slightly filled in, Nelson Irvine checked available Hα plates and found that in almost every case Hα was in emission. So I think that I was able to detect these marginal cases quite reliably.

Williams: Did you find any S stars?

Houk: Yes, four; however a number of other earlier and/or underexposed ones were undoubtedly classified as M. They are not very conspicuous on our plates except at the latest types.

THE BONNER SPECTRAL ATLAS AND
THREE-DIMENSIONAL CLASSIFICATION AT LOW DISPERSIONS

W. C. SEITTER

Bonn Observatory, Federal Republic of Germany and Smith College, U.S.A.

Abstract. Work on the third part of the *Bonner Spectral Atlas: Peculiar Stars* has well progressed during the past year. Observations of the more than 200 stars — photographed with a dispersion of 240 Å mm^{-1} at Hγ on *I-N* plates — is nearing completion.

The arrangement of the spectra will be as follows:
1. WR-stars
2. O-stars Of sequence
3. Peculiar B-type stars
   emission-line objects
4. Ap-stars
   with various sequences:
   Cr-Mn-Hg-rare earths
5. Asi-stars
6. Am-stars
7. Late-type peculiar stars Ba II, CH
8. C-stars
9. Late M-type stars
10. S-stars
11. Composite spectra
12. Spectra with large rotational broadening

The 12 groups are displayed on 40 plates, each with 6–8 objects. Stars of groups 8 to 10 will be presented with different exposures in order to facilitate the discovery of faint objects.

Sample plates will be shown and discussed.

1. Introduction to the Bonner Spectral Atlas

The Bonner Spectral Atlas — Atlas for Objective Prism Spectra — consists of the published or projected parts as shown in Table I.

2. Visibility of Classification Features on Low-Dispersion Spectrograms

While planning the various parts of the atlas it came, of course, to our minds, whether it was at all sensible to reproduce spectra of a large variety of spectral types at low dispersions when it might not be possible to differentiate between them. We thus looked for an empirical rule to predict the visibility to certain classification features and found the following: *The equivalent width of the faintest visible feature in m*Å *is approximately equal to the dispersion in* Å *mm*$^{-1}$.

This heuristic rule is derived from observations over a wide range of dispersions

B. Hauck and P. C. Keenan (eds.), Abundance Effects in Classification, 129–133. All Rights Reserved.

TABLE I

Published and projected parts of the Bonner Spectral Atlas

No.	Volume	Stellar Types	Dispersions	Publication Date
1	Part I	MK standards	240 Å mm^{-1}	1970
2	Part II	MK standards	645, 1280 Å mm^{-1}	1975
3	Supplement I	MK standards text only	240, 645, 1280 Å mm^{-1}	1976
4	Part III	'peculiar' stars	240 Å mm^{-1}	(1977)
5	Part IV	'peculiar' stars	645, 1280 Å mm^{-1}	(1979)
6	Supplement II	'peculiar' stars text only	240, 645, 1280 Å mm^{-1}	(1979)
7	Part V	Nova Delphini	240 Å mm^{-1}	(1976)

The term 'peculiar' refers to all types not included in the MK system of 1953, including such normal objects as M-type stars later than M2 and Population II objects.

$(0.5$ Å mm$^{-1} - 1280$ Å mm$^{-1})$ and seems to hold reasonably well for both absorption and emission lines, provided that the line width is not much larger than the resolution of the spectrogram. Exposure effects, small fluctuations in resolution and line width and uncontrolled influences set the limits between which this holds true to roughly ± 0.3.

One requirement is essential in all cases of high (relative to the dispersion) accuracy classification: sufficient broadening of the spectra. Regardless of dispersion, the widening should not go below 0.5 mm. Best results are obtained for widening values between 0.7 and 1.5 mm.

In order to compare the results obtained with a variety of instruments equipped with objective prisms to those obtained with the Bonn Schmidt telescope, which has the fairly short focal length 1375 mm and thus — with seeing values of no more than $2''$ — a high resolution, one has to apply the following conversion factor for telescope X:

$$\frac{\beta_X \cdot T_X}{\beta_{Bonn} \cdot T_{Bonn}}$$

where β = seeing, and T = focal length of the telescope.

Figures 1 and 2 show the predicted visibility of certain features on the basis of their equivalent widths according to the above cited rule.

The ordinates represent equivalent widths in mÅ as well as dispersions in Å mm^{-1}. The three Bonn dispersions are indicated by long lines. The abscissae represent spectral types. Figure 1 shows the equivalent widths of a certain silicon and europium blend in different types of stars. It is immediately apparent that the two higher dispersions should easily reveal Si II and Eu II stars on the basis of the line strengths around $\lambda 4130$.

Figure 2 shows the absorption strengths around $\lambda 4180$ indicating that the feature separates Am stars from normal stars at all three dispersions of the Bonn Schmidt telescope. All symbols are explained in the diagrams.

The results of both diagrams have been checked on objective prism plates and are found to hold true.

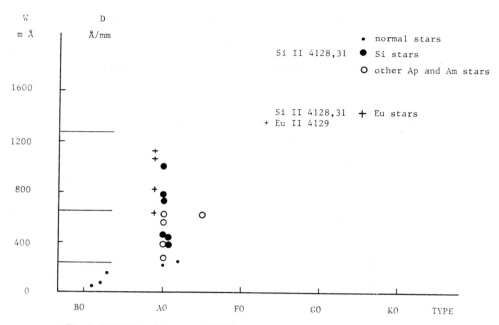

Fig. 1. Visibility of abnormal Si II line strengths at different dispersons.

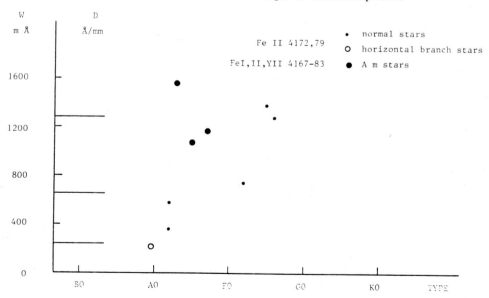

Fig. 2. Visibility of metallic star characteristics at different dispersions.

3. Sample Spectra of Different Peculiar Spectral Types
at Low Dispersions

A variety of spectra including Ap stars of the Si, Cr and Sr groups, Am stars, late M-type stars, C-stars, S-stars and emission-line objects at all three Bonn dispersions are shown on slides and in an exhibit in order to give the audience the opportunity to judge for themselves the visibility of features at the different dispersions.

The detectibilities of peculiarities of some major types are summarized in Table II.

TABLE II

Recognition of spectral peculiarities at different dispersions

| Stellar | Dispersion in Å mm^{-1} | | |
Type	240	645	1280
Of	+	+	(+)
Be	+	+	[+]
He-weak	+	[+]	[+]
A Si	+	+	[*]
A Cr	+	+	(*)
A Sr	+	[+]	−
A Eu	+	[+]	−
A Mn	[+]	[*]	[*]
A m	+	+	*
λ Boo	[+]	−	−
R CrB	+	+	+
Ba II	*	[*]	[*]
CH	+	*	(*)
CN weak	+	+	+
CN strong	+	*	*
C	+	+	+
S	+	−	−
composite	[+]	[+]	[+]

Characteristic features

+   clearly recognized
*   seen − but frequently confused with other peculiarities
−   not seen
() expected to show − not yet tested
[] only seen in exceptional cases

The above results are preliminary.

DISCUSSION

Morgan: What are the categories of 'normal' spectra most easy to recognize on your two lowest dispersions?

Seitter: At all spectral types the Ia stars are easily detected. These and other groups profit from the fact that at low dispersions new criteria appear in the form of characteristic blends and (quasi)-continuum features replacing the lines which are lost at low resolutions. An example is the broad blend of the Si IV 4089 4116 lines with Hδ, which constitutes an excellent luminosity criterion for hot stars at lowest dispersion. H-line strengths in O-F stars are very good combined spectral type-luminosity criteria.

The problem at low dispersions is not the *absence* of criteria, but the *ambiguity* of the criteria which are present. The use of as many criteria as possible helps to separate the different parameters T, g and abundances for more types than expected.

Walborn: I wonder if you can really resolve all of the two-dimensional categories of the MK system with this observational system, and if not, whether this classification should be called 'MK'.

Seitter: The problem of assigning MK types at low dispersions is comparable to the problem of assigning MK types at very high dispersions — in the latter case you have too much, in the former too little information. For the sake of uniformity I would vote for assigning MK types irrespective of dispersion. For the low dispersions this seems justified in view of the fact that there is *no break* in accuracy around dispersion 100 Å mm^{-1}. There is rather a linear decline between about 50 and 1500 Å mm^{-1} with [σ] increasing by 0.012 spectral subtypes per 100 Å mm^{-1} progress in dispersion and 0.08 luminosity classes per 100 Å mm^{-1} progress in dispersion as derived from Bonn spectra. Thus, if an MK type is given together with the dispersion used, the reliability of the type is immediately apparent.

THE SPECTRAL TYPE OF HD 101065

A. PRZYBYLSKI

Mount Strömlo and Siding Spring Observatory, Canberra, Australia

Abstract. The eighth magnitude star HD 101065 has an extremely peculiar spectrum dominated by numerous lines of the rare earths and lacking the lines of 'normal' elements, such as iron peak elements and lighter elements. For this reason it cannot be fitted into the adopted framework of spectral classification. The spectral type of this star can be defined only by fixing its effective temperature, which is 6040 ± 100 K. However, because of the extremely high blanketing the temperature of the continuum is about 450 K higher. This means that HD 101065 is a late F type star.

1. Introduction

There is probably no other star whose spectral classification has caused as much confusion as that of the eighth magnitude star HD 101065.

In the Henry Draper catalogue it is classified as a B5 star, which certainly it is not. On the basis of its *UBV* colours, it could be classified as a K0 star with an ultraviolet excess. On the other hand, from their six-colour photometry, Kron and Gordon (1961) classified it as an F8 star with the highest known blanketing effect. And finally, Wegner and Petford (1974) assigned the type F0 from their study of equilibria between neutral and singly ionized elements.

The real cause of this confusion ultimately lies in the abnormal chemical composition of the atmosphere of this star. Its spectrum lacks such 'normal' elements as magnesium, silicon, iron, chromium, titanium and other iron peak elements, while only traces of calcium are present. On the other hand the spectrum is entirely dominated by numerous lines of the rare earths. In addition, only the presence of strontium, yttrium, zirconium and barium could be established beyond any doubt.

About 3000 lines were measured in the blue region from λ 3650 to λ 4830 Å. In addition Wegner and Petford recorded about 2000 lines in the red region from λ 4806 Å on.

2. Effective Temperature

In view of the lack of normal elements in the atmosphere, HD 101065 cannot be fitted into the adopted scheme of spectral classification based on selected absorption features. For this reason an appropriate spectral class can be allotted to it only if its temperature is known and the effect of blanketing on the structure of its atmosphere can be evaluated.

The determination of the effective temperature was the subject of a paper now in press

B. Hauck and P. C. Keenan (eds.), Abundance Effects in Classification, 135–141. All Rights Reserved.
Copyright © 1976 by the IAU.

TABLE I

Six-colour photometry

Star	U	V	B	G	R	I
HD 101065[a]	+0.08	+0.06	−0.04	−0.02	+.07	+0.20
HD 101065[b]	−0.60	−0.30	−0.13	−0.03	+.16	+0.35
β Vir	−0.20	−0.17	−0.06	−0.03	+.09	+0.21
Procyon	−0.43	−0.38	−0.14	−0.03	+.17	+0.39

[a] observed colours

[b] colours corrected for blanketing

(Przybylski, 1975) and therefore only a few details are quoted here. As the six-colour photometry in Table I shows the radiation of HD 101065 between the blue (λ 4880 Å) and the infrared (λ 10300 Å) colours can be approximated quite well to the radiation of β Virginis for which Baschek *et al.* (1967) found an effective temperature of 6120 ± 100 K. In addition, in the far infrared the difference between both stars is still reasonably small. Both stars emit about 75% of their total radiation in the spectral region above the blue band (λ 4880 Å). However, they differ considerably in the amount of energy emitted in the violet and ultraviolet regions. Detailed calculations show that this difference amounts to about 5.7%. This difference is reduced to 5.2% by a small excess in the far infrared in the radiation of HD 101065. A difference of 5.2% in the emission means a difference of 1.3% or about 80 K in the temperature of both stars. Thus for HD 101065 we obtain a temperature of about 6040 K, which makes it an F8 star in full agreement with its classification by Kron and Gordon (1961).

3. Blanketing

In their paper on HD 101065 Kron and Gordon (1961) noticed that this star has the highest known blanketing for any late-type F star. This is obviously due to the numerous absorption lines of the rare earths. In fact, the whole spectrum between λ 3650 and λ 4830 Å can be considered as one large blend.

It is difficult to evaluate the blanketing effect since the position of the undisturbed continuum is problematic. In the present investigations the continuum between λ 3800 and λ 4800 Å was drawn through the highest intensity peaks. Above λ 4800 Å, where lines are weaker and less numerous, it is easier to fix the continuum but unfortunately the two available spectra taken on 130aF Kodak plates are underexposed below λ 5500 Å. The results obtained from both spectra do not agree well, and therefore errors up to about 20% are possible.

The results of the measurements are shown in Figure 1. No evaluation below λ 3800 Å was undertaken since there the 'true' continuum cannot be drawn at all. Only a crude geometrical extrapolation of the blanketing below this limit is possible. Extra-

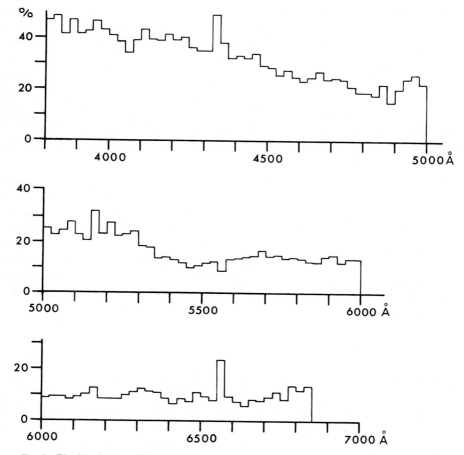

Fig. 1. The blanketing of HD 101065.

polation must also be used above λ 6850 Å since no good quality high-dispersion spectra of this region are available. In the present investigations it was assumed that 54% of the radiation is absorbed in spectral lines at λ 3550 Å, and 75% at λ 2000 Å and below this limit. Further, absorption of 6.8%, 5%, 3.4% and 2% was adopted for λλ 7000, 7500, 8000 and 8500 Å respectively. The assumption was also made that in the infrared band at λ 10300 Å the absorption falls to 1%.

With the help of Figure 1 we easily find that the absorption in V, B and G bands of six-colour photometry is 38.2, 20.3 and 14.6% respectively. For the U,R,I bands the corresponding figures found from the adopted extrapolation are 54, 6 and 1%. In terms of stellar magnitudes the corrections due to blanketing are then 0.84, 0.52, 0.25, 0.17, 0.07 and 0.01 mag. in order of increasing wavelengths. Subtracting those corrections from the observed six-colour photometry and adding uniformly a correction of 0.16 for the sake of normalization we obtain the de-blanketed colours shown in Table I. They agree tolerably well with the observed colours of *blanketed* Procyon (HD 61421), which has an effective temperature of 6500 K according to Strom and Kurucz (1969) and also to Carbon and Gingerich (1969). The temperature of deblanketed Procyon is about 150 K higher.

Judging from the data in Table I the temperature of the continuum of HD 101065 should be slightly lower than the effective temperature of Procyon, but the difference is certainly so small that it can be disregarded. A model stellar atmosphere with an effective temperature of 6500° can, therefore, be used for the numerical computation of the blocking coefficient η. The model of Carbon and Gingerich (1969) for a main-sequence star (log g = 4.0) was used for this purpose.

The total energy radiation output for this model is 32.215×10^9 erg cm^{-2} s^{-1}. Detailed computation in the region from λ 3800 – λ 6850 Å based on data from Figure 1 show that 10.86% of radiation is removed from the spectrum by spectral lines in this interval. The absorption below λ 3800 Å (computed with the help of the adopted, extrapolated blanketing down to λ 1683 Å) amounts to 13.65% of the total energy output of the star. The energy absorbed above λ6850 is only 0.78%. Thus the total amount of energy removed from the spectrum is 25.29% or 8.147 erg cm^{-2} s^{-1}. The blocking coefficient η is 0.2529. The effective temperature T_e can now be computed from the temperature T_o of the continuum with the help of the formula

$$(T_e/T_0)^4 = 1 - \eta.$$

Numerically we obtain T_e = 6043 K which means that the effective temperature is 457 K lower than the temperature of the continuum. This result agrees well with the temperature obtained from the comparison of HD 101065 with β Virginis but obviously this agreement is coincidental, since the accuracy of our calculations is rather low — more than half of the blanketing had to be computed from extrapolated estimates.

4. Comparison with the Results of Hyland

HD 101065 was recently observed by Hyland *et al.* (1975) in the far infrared colours J (1.25μ), H (1.65 μ) and K (2.2μ). They derived the temperature of the star from deblanketed colours $V -J$, $V -H$ and $V -K$, where V = 8.02 is the visual magnitude at λ = 5400 Å. The blanketing correction at λ = 5400 Å derived from Figure 1 is 0.18 mag. From their observations Hyland and his co-workers conclude that the temperature of the continuum of HD 101065 is essentially the same as that of F5 to F6 main-sequence stars with an effective temperature between 6300 and 6500 K. Adopting a differential correction of 120 K due to blanketing effect, they conclude that the effective temperature of HD 101065 is 6300 ± 150 K.

However, their differential correction seems to be too small. For an average F5-F6 star a mean effective temperature of 6400 K can be adopted. From Wildey's *et al.* (1962)

estimates of blocking coefficients η for several stars we can conclude that the blanketing correction for such a star is about 120 K, and thus we obtain $T_0 = 652C$ K for an average F5-F6 star. Adopting the same temperature of the continuum for HD 101065 and applying a blanketing correction of 457 K from the present investigations, we obtain an effective temperature $T_e = 6063°$, not much different from comparison of the star with β Virginis and with Procyon. Hyland's observations can, therefore, be well reconciled with the present results.

5. Final Remarks

Objective prism spectra of low dispersion, such as for instance spectra used for the Henry Draper classification, cannot record thousands of narrow spectral lines seen on high-dispersion spectra of HD 101065. For this reason almost all details are lost and only a few broad features can be seen. Among the few visible features are obviously the hydrogen lines. From their strength the star could be classified as a late F or a B5 star. Apparently because of the lack of any visible ionized calcium lines H and K, the B5 type was chosen for the Henry Draper catalogue. Two objective prism spectra of 470 Å mm^{-1} dispersion at Hγ taken in 1961 by B. Westerlund and G. Lyngå with the 26″ Schmidt telescope of the Uppsala Southern Station show that the spectrum of HD 101065 really looks like that of a B5 star with a wrong continuum and without helium lines. Therefore it can be safely assumed that the star is not a spectrum variable in spite of its classification as a B5 star in the Henry Draper catalogue.

In Johnson's three-colour photometry ($V = 8.017 \pm 0.004$, $B-V = 0.763 \pm 0.003$, $U-B = 0.241 \pm 0.006$) HD 101065 seems to be a K0 star with an ultraviolet excess of 0.22 mag. The high $B-V$ value is obviously due to high blanketing in the violet band ($\lambda 4220$ Å) which more or less corresponds to Johnson's blue colour. In normal F stars the blanketing amounts to about 15% while in HD 101065 it is $2\frac{1}{2}$ times higher (38.2%). This causes a relative difference of 0.34 mag. in the $B-V$ colour — only partly compensated by increased blanketing in Johnson's visual band.

If we adopt Milne's law of temperature distribution in the atmosphere and assume that spectral lines are formed by true absorption, the maximum limit for blanketing is 50% of the undisturbed radiation. In real stars the boundary temperature drops below the limit imposed by Milne's law and, in addition, spectral lines are formed not only by true absorption but also by the process of scattering and, therefore, more than 50% of energy can be removed in spectral lines. In fact, the present investigations show that at $\lambda 3800$ Å almost 50% of the radiation is removed and certainly more than this is removed in the ultraviolet band. Obviously, however, the increase in blanketing is becoming more difficult when already a large amount of energy is removed from the spectrum. In normal late-type F stars about 35% of radiation is removed in the ultraviolet band ($\lambda 3520$ Å). In HD 101065 the extrapolated degree of absorption is 54%, or 19% more than in F stars. In terms of stellar magnitudes this difference is equal to 0.37 mag. or only slightly more than in Johnson's blue band. Thus the $U-B$ difference remains practically unaffected. HD 101065 has, therefore, the $B-V$ colour of a K0 star but the $U-B$ colour of a late-type F star. This explains the apparent ultraviolet excess of the star in Johnson's system.

An increase of metal abundance in the star would probably increase this ultraviolet excess. It would cause a relatively small increase of blanketing in the visual region, a much larger increase in the blue band and a relatively small increase in the ultraviolet. As a result we would obtain an increase of the $B-V$ colour but a decrease of the $U-B$ colour. In this way the ultraviolet excess would be increased. In astronomical practice the ultraviolet excess is associated with the underabundance of metals, and conversely the ultraviolet deficiency with the overabundance of metals. HD 101065 is an exception to this general rule because of its enormous blanketing.

The total abundance of metals, whose presence in HD 101065 was established in an analysis of the spectrum (Przybylski, 1966), possibly does not exceed 1/100 of the abundance of iron in normal stars. The star may thus be metal poor in spite of its large blanketing effect. However, this is not a firmly established fact, since additional elements may still be discovered in the spectrum and since the abundance of elements not represented by lines in the spectra of late-type stars is an unknown factor in estimations of the total metal abundance. Recent investigations show that two elements, yttrium and zirconium must be added to the list of elements present in the star. On the other hand the presence of several elements reported by Wegner and Petford (1974) is doubtful. At present this is the subject of a controversy, which will hopefully be resolved in the near future.

The position of the star in the Russel-Herzsprung diagram is unknown. Unfortunately the parallax found by Churms ($\pi = 0\rlap{.}''004 \pm 0.006$) is small and its probable error exceeds the value of the parallax itself. Hardly any conclusion can be drawn from such results. An eighth magnitude F8 star should have a parallax of $0\rlap{.}''016$ in order to lie on the main sequence.

Acknowledgements

The author wishes to acknowledge with thanks the help of Mrs Dagmar Gilfelt who measured the blanketing of HD 101065 and of Mr Bela Bodor who prepared Figure 1 for publication.

References

Baschek, B., Holweger, H., Namba, O., and Traving, G.: 1967, *Z. Astrophys.* **65**, 418.
Carbon, D. F. and Gingerich, O.: 1969, *Proceedings of the Third Harvard Conference on Stellar Atmospheres*, Cambridge, Mass., p. 377.
Hyland, A. R., Mould, J. R., Robinson, G., and Thomas, J. A.: 1975, *Publ. Astron. Soc. Pacific* (in press).

Kron, G. E. and Gordon, K. C.: 1961, *Publ. Astron. Soc. Pacific* **73**, 267.
Przybylski, A.: 1966, *Nature* **210**, 20.
Przybylski, A.: 1975, *Monthly Notices Roy. Astron. Soc.* (in press).
Strom, S. E. and Kurucz, R. L.: 1966, *J. Quantit. Spectrosc. Radiative Transfer* **6**, 591.
Wegner, G. and Petford, A. D.: 1974, *Monthly Notices Roy. Astron. Soc.* **168**, 557.
Wildey, R. L., Burbidge, E. M., Sandage, A. R., and Burbidge, G. R.: 1962, *Astrophys. J.* **135**, 94.

DISCUSSION

Spinrad: Now that you have the temperature, what is the Ho overabundance in the star?

Przybylski: All the rare earths are overabundant by a factor of about 1000 if we assume that the continuous opacity is normal. Unfortunately we cannot be sure of that.

CLASSIFICATION OF STARS WITH DIFFERENT
CHEMICAL CONSTITUTION

D. CHALONGE and L. DIVAN

Institut d'Astrophysique, Paris, France

Abstract. The possibility of obtaining a segregation between stars of different chemical constitution using the BCD classification is reviewed.

It is now well known that large differences in chemical composition do exist between the various families of stars, but the MK classification seems to be reserved to the so-called 'normal stars' having a chemical composition more or less similar to the solar one: neither the 'metallic line stars' nor the stars of population II which have chemical compositions very different, and very different from the one of the normal stars, can find a place in it.

After the pioneer work of Unsöld on τ Scorpii, the accurate analysis of the composition of a number of normal stars, performed in various observatories, has shown that rather small but nevertheless conspicuous differences of chemical constitution exist between them.

Thus, since large differences of constitution prevent a classification, one may wonder whether the small differences observed between the normal stars do not bring certain distorsions in their classification.

The conclusion is that it is desirable to improve the actual system of classification in order to open it to all kinds of chemical composition. This would require the introduction of, at least, one more parameter to take into account the behaviour of stars with different compositions.

During a symposium on spectral classification held in Paris in 1953, we described such a three-dimensional system of stellar classification and gave its first results: they seem to be in good agreement with those of the MK classification for the normal stars and proved that this new system was opened to stars with composition so different as the metallic line stars and the subdwarfs.

Since that time we have applied the method to the classification of several hundreds of stars of different types in order to give a stronger foundation to this new system, and we have now enough material to confirm and to complete our first conclusions. Some of these new results have been presented during *Symposium* no. 24 held in 1966 at Saltsjöbaden (the corresponding paper will be referred to as number I), and, with more details, last year in a paper published in *Astron. Astrophys.* (which will be referred to as number II).

So we give here only a short account on the method and on the actual situation. In this system of classification every star is characterized by the numerical values of three quantitative parameters describing the distribution of energy in the continuum of the star. Our parameters concern a layer of the star somewhat deeper than the layers used in the MK classification but unfortunately this method requires, for the measurement of the values of the parameters, that the star to be classified presents, all along its spectrum, 'windows' permitting to reach these deeper layers.

This condition is fulfilled only in the early types, O, B, A, F, so that our classification is restricted to these types, that is, to the more massive and younger stars. The three parameters, all pertaining to the continuum, are the spectrophotometric gradient ϕ_b corresponding to the colour temperature of the continuum in the blue-violet region of the spectrum, the Balmer discontinuity D and a wavelength λ_1 characteristic of the real position of the discontinuity in the spectrum.

The first parameter is very sensitive to the interstellar absorption, so that its real value can be directly determined only for the nearby stars.

But the two other parameters are not affected by this absorption: they give, without any correction, the value of two intrinsic features of the star and the main advantages of the present system of classification are direct consequences from this fact. When we know the intrinsic values of the three parameters for a star we may represent it by a point with the co-ordinates ϕ_b, D and λ_1 in a three-dimensional system of co-ordinates. The representative points of the normal stars fall very near to a surface which is represented in Figure 1 (paper I): the different spectral types are distributed in well-determined areas on the surface. Figures 2 and 3 of paper I show respectively that the metallic-line stars and the population II stars, which have a different chemical composition, are represented by points situated well out of the surface and on both sides of it. The few blue stars of the halo represented on Figure 3 (paper I), which probably belong also to population II, would eventually not lie far from the surface if their gradient was corrected from a small interstellar absorption.

If we return to the normal stars, those who have been found somewhat more metal rich or more metal poor lie in the immediate vicinity of the surface, on the right side for the first, on the left for the second.

Figure 1 of the present paper represents the same results in another system of co-ordinates: every star is represented by a point with ϕ_b as ordinate, and as abscissa a variable s giving a quantitative measure of the spectral type; the definition of s has been given in paper II. The few not reddened O and B stars are distributed along a very regular line but the late A and the F stars of Figure 1, which are all nearby stars with no interstellar reddening, show a significant scatter and lie between the two dotted lines. An average line has been drawn between them and it may be seen that the metal-rich star β Vir falls above the average line, and the metal-poor χ Dra, ξ Peg, σ Boo fall below. The majority of the Hyades are situated above the average line and also the members of the galactic cluster in Coma: this suggests that they are somewhat more metal-rich than the Sun, a result which has been found by various observers for later members of the two clusters. The average line would correspond to stars having exactly the same composition as the Sun.

The difference in the distribution of the B and of the F stars is probably due to the

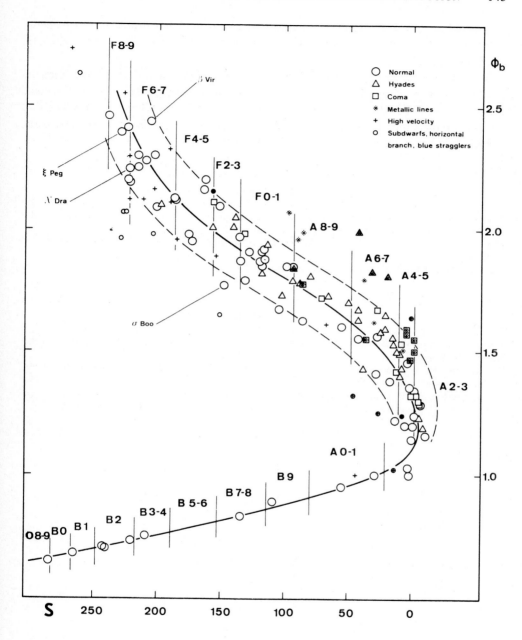

fact that the F belong (because of their relatively slow evolution) to several generations of stars with various compositions and the B only to the last one, a rapid evolution having brought the others to other parts of the diagram.

The metallic-line stars, fall in general more above the average line than the above-mentioned metal-rich stars and the subdwarfs and horizontal branch stars lie much below this line; the high velocity stars lie also below the average line; a part of them belong to the population II.

It thus appears that our classification agrees with the MK one only for stars lying on the 'average line' of Figure 1; and these stars should be considered as the real 'normal stars'. For the others the figure suggests that they are either somewhat more metal-rich or metal-poor than the normal stars, a result which cannot be expressed by the MK parameters alone and would require the use of a third parameter.

References

Chalonge, D.: 1955, *Principes Fondamentaux de Classification Stellaire,* Colloques Internationaux du Centre National de la Recherche Scientifique, Paris, p. 55.

Chalonge, D.: 1966, in K. Lodén, L. O. Lodén and U. Sinnerstad (eds.), *IAU Symp.* **24**, 77, Academic Press, London.

Chalonge, D. and Divan, L.: 1973, *Astron. Astrophys.* **23**, 69.

DISCUSSION

Cayrel: Are the iso-metal-content lines you found for the Hyades and Coma the same as those in the Strömgren photometry?

Chalonge: We have not studied the same members of the two clusters as those observed in the Strömgren photometry: we have considered their earliest members and, in the Strömgren spectroscopy, latest members have been observed But both works lead to the same conclusions: these two clusters are somewhat more metal-rich than the Sun. We have not given iso-metal-content lines.

EFFECTS OF HEAVY-ELEMENT ABUNDANCE ON CLASSIFICATION OF G-TYPE GIANTS

P. C. KEENAN

Perkins Observatory, Ohio State and Ohio Wesleyan Universities, U.S.A.

Abstract. Five stars, including ζ Cyg, which had been classified as G6 II to G8 II bright giants, were found by O. C. Wilson to have K-line luminosities close to those of class III giants. These stars show enhancement of the lines of Ba II and other heavy-metal ions also. To eliminate the serious effect of this abundance anomaly on the spectroscopic luminosities new luminosity criteria involving only intercomparisons of lines of elements of the fourth period (Ti, Fe, etc.) were applied on 9 Å mm^{-1} Coudé spectrograms taken by O. C. Wilson. This Coudé classification gave luminosity classes near IIIa for these stars, implying absolute magnitudes considerably below those of bright giants but somewhat above Wilson's M_K values.

Another advantage of Coudé classification is the possibility of estimating luminosities for individual barium stars. From Wilson's plate of HD 205011 a luminosity class of III-IIIa is derived. This is consistent with the mean value of $M_v = -0.4$ derived from statistical parallaxes by MacConnell, Frye and Upgren (*Astron. J.* 77, (1972), 384) for the barium stars.

The detailed account of this investigation will be published elsewhere.

DISCUSSION

Williams: Do you have any comment on the luminosity of the prototype BaII star ζ Cap, which appears from Wilson's M_v (K) to be very luminous?

Keenan: I have no spectrograms of this star, and have no reason to think that it may not be as bright as $M_v \approx -3$.

SPECTRAL CLASSIFICATION IN THE SMALL MAGELLANIC CLOUD

P. DUBOIS, M. JASCHEK, and C. JASCHEK

Observatoire de Strasbourg, France

Abstract. This paper presents results obtained in classifying 26 supergiants of the Small Magellanic Cloud. The main results are:

– The hydrogen lines are not as sharp as expected for such luminous stars.

– Generally, for the A-type stars, the calcium K-line is too strong compared to the other metallic lines.

– There are differences in the behaviour of the different metals.

– Often, the stars present variations in their spectra.

– 50% of the stars brighter than $m_v = 11.3$ have an enhancement of the H8 Balmer line, which is one of the characteristics of a shell star.

– The comparison with the Radcliffe classification and Dachs classification is, except in very few cases, in good agreement.

It is well known that there are problems connected with the chemical composition of the SMC stars, the opinions ranging from solar composition to various degrees of under-abundance of metals. Because chemical composition analyses require high dispersion and these in turn very long exposure times, we have adopted a simpler method. We have tried to compare a sufficiently large number of SMC supergiants to galactic supergiants. This paper gives some results concerning 26 B and A-type SMC supergiants.

The objects were selected from the list of Florsch (1972), who gives the radial velocity measured with Fehrenbach's objective prism, and from the list of Sanduleak (1968). Most of the stars are common to both lists. The sample of galactic supergiants is composed of 38 stars classified by Morgan.

The spectra were obtained at the ESO Observatory in Chile with a slit grating spectrograph at the Cassegrain focus (RV Cass) which gives a dispersion of 74 Å mm^{-1}. The projected slit width on the film is 16 μ (= 1.2 Å). Because of the exposure times, the spectra have to be trailed in height only over 200 μ. The spectra show the whole region from λ 3600–4900 Å. Some spectra taken by Azzopardi with the same instrument were kindly lent by him for this work.

The spatial distribution of the SMC stars is the following: 3 are located in the wing, 20 in the bar, mainly at the two ends, and 3 lie outside. Among them we have 8 B-type stars and 18 A-type. The proportionally larger number of A-type stars comes from the fact that we chose them from the list of Florsch. For the A-type stars the Balmer line have the largest contrast and are thus more easy to measure by the objective prism technique.

When classifying these stars one runs immediately into problems, because the stars do definitely not show spectra like those of their galactic counterparts.

The first characteristic is that the hydrogen lines are not so sharp as could be expected of such high luminosity stars, except in a few cases. One finds stars having an absolute magnitude of about −7 which have hydrogen lines broader than those of galactic super-giants of luminosity class lb, to which we assign generally an absolute visual magnitude of about −5.

As a result of this, the number of the last visible Balmer line is not larger than in galactic bright supergiants of similar spectral type. Furthermore, the Balmer jump in many cases corresponds to an earlier spectral type.

The line broadening described is not restricted to the hydrogen lines, but is shared in general by all the other lines.

A second striking characteristic is, in the A-stars, the absence or extreme weakness of the metallic lines. A closer analysis reveals however that this weakening is not uniform for all metals. In particular one finds in the A- stars, if one designs by $S(X)$ the spectral type obtained from the lines of element X, that:

$$S(Ca) \gtrsim S \text{ (Ti II)} > S \text{ (Fe II)}$$

Not only is Fe II weaker than expected, but, moreover, Fe I lines are almost never visible.

Si II and Mg II lines are often — but not consistently — weaker than Ti II. The difference between S (Ti II) and S (Fe II) is of the order of 2 tenths of spectral type, a fact which is easily visible in the interval A0-A2.

In the B stars, the weakening of elements other than hydrogen and helium is harder to detect. However, one does not see in the stars taken any enhancement of C, N or O as could be expected in view of the luminosity of these stars. It should be added that perhaps also the Si is weaker than expected. For two of them Osmer (1973) found weak lines of N, O, Si and probably Mg and C.

The third important point is that the Ca II lines are present not only in the A-type stars, but also in the B-type stars, except in two stars. But in both of our two exceptions, Feast *et al.* (1960) found on their spectra a K-line. The radial velocity of the K-line for all the A- and B-type stars coincides with the one determined from the other lines, within the errors of measurements.

A fourth characteristic is the spectrum variability. Since we have up to five spectra for each star, this becomes easily observable. The variations occur over time scales of one or several days. This kind of activity of the supergiants is well known from the studies of Rosendahl (1973), Hutching (1971, 1973) and others. Looking for a common characteristic of the stars which do vary, one finds that 5 out of 7 stars, earlier than B5, show at least on one plate an enhancement of the H8 line, which is a well-known characteristic of shell stars. However, not all other characteristics of shell stars — sharpness of lines, underlying broad lines, presence of numerous metallic lines — are always present. The enhancement of H8 occurs even in two A-type stars, so that if we consider all stars brighter than $11^{m}.2$, which corresponds to about $M_v \simeq -8.3$ if one puts all cloud stars at the same distance, one finds that 50% of the stars do show this characteristic.

The existence of spectrum variability explains at least part of the differences found when comparing the present classification with those of Feast *et al.* (1960) and with

Dachs (1970). There exists a systematic effect with the Radcliffe classification in the sense that our classifications are about one tenth later; the remaining dispersion is 1.5 tenth. The systematic effect is probably due to the fact that this classification relies more upon the intensity of the K-line than does that of Feast *et al.* (1960). The spectral types given by Dachs are based upon photometric measurements which depend mainly upon the Balmer discontinuity. Their agreement with the present work is poorer; the dispersion is 2.5 tenth, due probably to the fact that one compares classifications based upon different criteria.

Summarizing, it can be said that we have a certain number of facts which can only be interpreted when good atmospheric models of very bright supergiants become available. Since these do not yet exist, one can only conclude that one does not know if the anomalies found are due to a deficiency in metals or an abnormal atmospheric structure which enhances certain metals preferentially over others.

References

Dachs, I.: 1970, *Astron. Astrophys.* 9, 95.

Feast, M. N., Thackeray, A. D., and Wesselink, A. J.: 1960, *Monthly Notices Roy. Astron. Soc.* 121, 337.

Florsch, A.: 1972, Publication de l'Observatoire Astronomique de Strasbourg, Vol. 2 fasc. 1.

Hutching, J. B.: 1971, in M. Hack (ed.), *Colloquium on Supergiant Stars,* Osservatorio Astronomico di Trieste, Trieste, p. 38.

Hutching, J. B. and Laskarides, P. G.: 1973, *Publ. Dominion Astrophys. Obs.* 14, 107.

Osmer, P. S.: 1973, *Astrophys. J.* 184, 127.

Rosendhal, J. D.: 1973, *Astrophys. J.* 186, 909.

Sanduleak, N.: 1968, *Astron. J.* 73, 246.

DISCUSSION

Przybylski: You find that the K-line of Ca II is too strong. I think one can generalize this statement by saying that lines arising from the ground level have the tendency to be too strong. This may be the reason why in the investigation of the brightest LMC supergiant, HD 33579, by Dr B. Wolf and me, we found an overabundance for aluminium.

Bidelman: I believe it is possible that the abnormally strong K-line that you find is simply due to interstellar calcium, as it is known that the small Magellanic Cloud has a large amount of gas.

Dubois: I can only comment that in view of what I said it looks doubtful that this is the real explanation.

Cayrel: Could you tell me your personal opinion about a possible metal/hydrogen deficiency in the SMC?

Dubois: Probably the best answer is contained in the last paragraph of my talk.

Cayrel: Who found that the SMC is metal poor?

Przybylski: I found that the brightest star, HD 7583, is metal poor by a factor of 10. However, Dr B. Wolf found normal abundances in this star. The discrepancy in the results is (at least) partly due to the difference in fixing the level of the continuum. My continuum cuts the photographic noise so that half of it is above and half below the adopted continuum. To the best of my knowledge Dr Wolf puts the continuum higher.

Bell: Parsons and I believe the SMC cepheids may be metal deficient by a factor of ~4 in view of their blue colour relative to galactic cepheids.

Bell: (to Przybylski): Why don't you use your calibration spectra to estimate grain noise instead of stellar spectra in which weak lines may be present?

Przybylski: I did some investigations of that kind. I concluded that in the investigation of narrow spectra (of, say, 0.3 mm width) the photographic noise can introduce spurious lines of the order of about 7 mÅ if I remember well, provided that we fix the continuum in the way described in my answer to Mrs Cayrel's question. However, the strength of the spurious lines may be five times larger if we draw the continuum through the highest peaks. In the investigation of stellar spectra we can expect similar errors.

The photographic noice is of no importance if we investigate broad spectra. Unfortunately in spectroscopic investigations there is no substitute for good spectra.

Walborn: (1) The lines could appear broader if the densities of the SMC spectra were systematically lower than those of the bright standards.

(2) I would be cautious about interpreting as real, variations of weak features on narrow, low-dispersion spectrograms.

(3) I have obtained 78 Å mm^{-1} 1.2 mm wide spectrograms of two B0 supergiants in the SMC; the Si IV lines have strengths similar to those in galactic standards of luminosity classes IV and V, and the carbon and nitrogen features are very weak. These results are in agreement with those of Dr P. S. Osmer on metal deficiency in SMC supergiants.

Dubois: With the first point I agree, and for this I have taken many spectra of each comparison star with various densities to eliminate this direct.

To the second point, some variations are very obvious, so that the reality cannot be questioned.

The third point is in agreement with what we have observed.

McCarthy: Can you explain fully how your estimate of the Balmer jump influenced your spectral classification of these stars? You mentioned that the Balmer jump corresponds to an earlier spectral type for several SMC stars.

Dubois: For 'A' stars the Ca K line is the chief criterion adjusted later by reference to Balmer jump.

Wolf: (1) A possible reason for some enhancement of the calcium K-line in the A-type stars could be the existence of systematic depth-dependent velocity fields in these extended atmospheres. In the case of HD 33579 [A2Ia-O] in the LMC, I have found the calcium H- and K-line to be asymmetric, thus providing direct observational evidence for the existence of some systematic velocity-field.

(2) I carried out a model atmosphere analysis of HD 7583 in the SMC, according to a similar technique that Groth (1961) applied to analyse α Cyg. As a result the metal abundance came out to be more likely solar within a factor of two, than considerably underabundant. But non-LTE effects, deviations from plane-parallelism, and systematic photospheric velocity fields, not considered in the analysis, are important in these extended atmospheres. These effects may influence the abundance determination and could be a physical 'explanation' for the supersonic 'microturbulence' determined in the analysis as well.

NITROGEN AND CARBON ANOMALIES IN OB SPECTRA

N. R. WALBORN

Cerro Tololo Inter-American Observatory, La Serena, Chile

Abstract. The following new OBN and OBC stars have been found: HD 75860, BC 2 Iab; HD 104565, OC 9.7 Ia; HD 123008, ON 9.7 Iab; and HD 150574, ON 9 III(n). Examples of morphologically more moderate CNO anomalies in OB spectra are also discussed; it is concluded that the O9-B0 7 supergiants of the Orion Belt and NGC 6231 (the nuclear cluster of Scorpius OB1) are systematically nitrogen-deficient. One *possible* explanation of these nitrogen-deficient supergiants in sub-associations is that they may have formed from nitrogen-deficient clouds; another could be that the more numerous morphologically normal supergiants show the effects of mixing in their spectra, while the nitrogen-deficient ones are relatively younger or have not yet been red supergiants. No red supergiants are found in Sco OB1, whereas NGC 3293 contains two B0 supergiants with normal CNO spectra together with a red supergiant. The paper will appear in the *Astrophysical Journal,* April 15, 1976.

DISCUSSION

Garrison: Your system depends very heavily on ratios of the He with Si. How confident are you that there are no Si anomalies and that all differences are due to CNO anomalies?

Walborn: The *spectral types* depend mostly upon ratios of successive stages of ionization of He and Si. The *luminosity classes* depend upon ratios of Si to He; however, the whole spectrum is inspected for consistency. For instance, if Si/He indicated a supergiant class, but the Balmer lines had wings, a discrepancy would be obvious; such cases are not seen. As in the classical MK system, the classification is an empirical procedure, whose validity is checked later by the calibrations. The He-Si types agree systematically with MK types for stars in common, and the calibration results are satisfactory.

Houziaux: It is remarkable that such a refined classification can be obtained for very hot objects since in your spectral range you observe, you collect a very small fraction of the total energy of the star. I am wondering if it would not be wise to wait now for satellite UV data, as the spectral differences should be much more spectacular in this wavelength range, the Planck function being much more sensitive to differences in temperature between 1200 and 2000 Å than it is in the 3800-5000 Å region.

Walborn: The absolute-magnitude and effective-temperature calibrations support the validity of the classification in the blue-violet spectral region. The more highly refined this classification can be made, the more valuable will be the eventual comparison with far-UV results.

INFRARED LINES IN PECULIAR EMISSION-LINE STARS

Y. ANDRILLAT

Observatoire de Haute-Provence (CNRS), France

and

L. HOUZIAUX

Université de Mons, Département d'Astrophysique, Belgium

Abstract. An image-tube Cassegrain spectrograph, equipped with an S-1 photocathode enabled us to obtain infrared spectra of peculiar emission-line objects, as V 1016 Cyg, HBV 475, HD 51585, HD 45677, XX Oph and CI Cyg. Many of these objects display emission lines of H I, He I, He II, O I, N I, Ca II [S II] [S III], [Fe II]. The line intensities can be used together with data from other parts of the spectrum to determine relative abundances of these ions in the shells of these objects, which may be at a critical phase of their evolution.

The image-tube spectrograph installed at the Cassegrain focus of the 193 cm telescope of the Haute-Provence Observatory enabled us to obtain infrared spectra of peculiar emission-line stars. The S-1 photocathode permitted to cover the range 8000−11 000 Å with a dispersion of 230 Å mm^{-1}. In addition, IN plates have been used to cover the 5800−8800 Å range with the same dispersion.

The peculiar objects include HD 51585, V 1016 Cyg, HBV 475, HD 45677, and XX

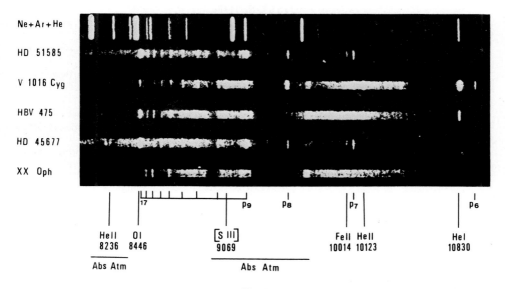

Fig. 1.

B. Hauck and P. C. Keenan (eds.), Abundance Effects in Classification, 155−156. *All Rights Reserved.*

Oph. In addition to light and spectral variability, these stars exhibit (a) characteristic continua of hot stars, (b) characteristic continua of cool stars, (c) infrared excess (3-10 μ region), (d) stellar and nebular emission and/or absorption lines, (e) molecular bands in absorption, (f) abnormal radio emission. It has been suggested that they are double stars including a long period variable ejecting mass, and a hot companion responsible for the excitation of the nebular spectrum.

HD 51585 shows a well-developed Paschen series (P6 to P23), together with O I, He I, N I, Ca II and [Fe II] lines. Two features at 9180–9205 Å are present unidentified.

V 1016 Cygni reveals lines of H I, He I, He II, [S II], [S III], O I, but Ca II is absent.

HBV 475 (a radio star) has a variable spectrum where [S III], He II, He I, Ca II, O I, and H I lines are detected. Wide features identified in the visible to He II, C III, N III, N IV transitions indicate that this object is close to a WR type star.

The moderate excitation object HD 45677 is similar to HD 51585, except for the two unidentified features, which are absent.

XX Oph, which shows so many bright lines in the visible, is rather poor in the infrared, except for O I and Ca II emissions. The strong P Cyg profile lines He I 10830 and P 6 observed in 1974 have disappeared on the 1975 spectrogram.

These spectra will be used for relative abundance determination using neighbouring lines: He/H (with He I 10830/P 6 and He II 10123/P 7 ratios), O/H (O I) 8446/P 17 ratio), N/H ([N II]/Hα and N I 8630/P 13 ratios) and S/H ([S III] λ9069/P 9 ratio). It is interesting to compare these ratios to the ones derived for planetary nebulae, since the nebulae emitted may be associated with H and He burning phenomena in shell sources.

DISCUSSION

Morgan: Did you observe spectral variations in the old nova T Cr B?

Houziaux: Our observations of this object in the infrared cover only a short period of time over which we did not, however, observe any variations.

McCarthy: For V 1016 Cyg can you tell me the source of emission features to the shortward of the He 10 830 Å emission feature (near 1 μ) and whether you have observed any variations in these features?

Y. Andrillat: [S II] 10 284 Å est présent dans V1016 Cyg. En ce qui concerne les variations, il faut être prudent car, pour les étoiles faibles, obtenues avec un temps de pose assez long, les raies de OH du ciel nocturne peuvent contaminer [S II], notamment la forte raie de OH à 10 273 Å.

Dans l'infrarouge, nous n'avons noté aucune variation dans TCrB car nos observations portent sur un très court laps de temps.

CONFLICTING EVIDENCE ON THE COMPOSITION OF AM STARS

CLAUDE VAN 'T VEER

Institut d'Astrophysique de Paris, France

and

CLAUDE BURKHART

Observatoire de Lyon, France

Discordance à propos de la composition d'étoiles Am

Résumé.

Les étoiles Am sont des objets typiques qui posent notamment des problèmes au niveau de la relation entre la classification et les abondances.

Nous avons étudié des étoiles Am de différentes températures et luminosités et quelques étoiles Am occupant une position particulière dans des diagrammes photométriques.

Nous avons trouvé pour certaines étoiles un désaccord entre les indices dits d'abondance et les abondances déterminées à grande dispersion.

Il s'agit souvent de géantes, ou de binaires spectroscopiques ou d'étoiles ayant une rotation observable. Cependant, il n'y a pas de relations entre ces différents phénomènes et les désaccords constatés. Nous donnerons quelques exemples d'étoiles que nous avons étudiées.

Les indices d'abondance sont-ils réellement dans tous les cas ce qu'ils sont supposés être? Dans quels cas ne le sont-ils pas et pourquoi? Ces questions restées sans réponse jusqu'à présent devraient être discutées.

Abstract. Our programme involves the study of Am stars of differing temperatures and luminosities as well as of the Am stars deviating from the positions of the classical ones in the photometric diagrams. We have found for certain stars a significant disagreement between the so-called 'abundance indices' and the abundances determined at high dispersion. Often giants or spectroscopic binaries, or stars with a detectable rotation, are involved. However, there are no consistant relationships between these different peculiarities and the observed disagreements. We shall give some examples of stars studied by us. Are these 'abundance indices' really what they are supposed to be in all cases? In which cases are they not and why?

1. Effets de la binarité sur les indices d'abondances

Nous avons observé une étoile à fort indice Δm_1, classée Am par Bidelman. Nous avons obtenu, à 12 Å mm^{-1}, un très beau spectre à doubles raies. Dans l'analyse détaillée à grande dispersion, par la méthode des courbes de croissances différentielles, avec les modèles d'atmosphères 'blanketés' de Carbon-Gingerich, l'étoile se révèle être normale. Nous pensons que la binarité est la raison principale du désaccord entre la classification photométrique et l'analyse détaillée d'abondances. En rassemblant les quelques étoiles Am (SB1, SB2) dont les abondances ont été déterminées, on s'aperçoit que la relation Δm_1/anomalies d'abondances présente de fortes contradictions.

B. Hauck and P. C. Keenan (eds.), Abundance Effects in Classification, 157–159. All Rights Reserved.

D'autre part, une estimation rapide de l'influence de la binarité sur les indices de Strömgren m_1 et c_1 a permis de mettre en évidence des effets non négligeables sur ces 2 indices. Les photométristes de l'Observatoire de Genève ont traité ce problème pour les indices du système de Genève. En première approximation nous pouvons dire que la binarité spectroscopique peut produire de faux effets d'abondance et de luminosité. Et en conséquence, on ne peut affirmer qu'une étoile est Am parce que $\Delta[m_1] \leqslant -0.025$.

Les photométries comme les classifications à faible dispersion ne sont pas toujours suffisantes pour conclure des anomalies d'abondances d'une étoile. Le problème est d'autant plus important pour les étoiles Am que, statistiquement, elles sont toutes binaires spectroscopiques.

2. Détermination de l'abondance en Ca des étoiles Am

Les figures $\Delta[m_1] / \Delta k$ de R. C. Henry (1969) et de R. C. Henry et J. E. Hesser (1971) montrent que le critère classique d'un type spectral K faible (comparé au type hydrogène) n'est pas capable de mettre en évidence le caractère métallique d'un certain nombre d'étoiles.

Il est intéressant de voir si les analyses spectrophotométriques fines trouvent que le calcium est normal pour ces étoiles Am. Les résultats pour 22 Boo (étoile toujours classée à raies métalliques, ayant un $\Delta[m_1]= -0.05$ comparable à celui de 63 Tau) sont: 22 Boo présente les anomalies d'abondance caractéristiques des étoiles Am; la déficience en Ca est faible mais réelle ($[Ca/H]^*_{\odot}= -0.45$ si on utilise les raies de Ca I ou les ailes observées de la raie K comparées aux profils théoriques calculés par Henry). Mais dans le diagramme $\Delta[m_1] / \Delta k$ de Henry, la raie K serait au contraire normale.

De même pour les autres étoiles étudiées dans le groupe Van 't Veer, nous avons des désaccords qui nous semblent significatifs. Mais plus les etoiles sont froides plus il y a de raies parasites mesurées dans les 2 filtres photométriques de Henry. Ce dernier en tient compte, mais suppose que les raies ionisées parasites se comportent comme la raie K de Ca II. Aussi, pour les étoiles Am qui ont plus de raies parasites qu'une étoile normale de même type K (ou H), (et même d'autant plus que l'étoile est plus 'métallique'), l'écart à la normale mesuré Δk ne traduit pas seulement une raie K faible (et donc une abondance faible en Ca), mais aussi l'exaltation des raies métalliques pour une température et une luminosité données.

Il sera important d'assurer ces résultats pour savoir:

— si les photométries $ubvy$, β et k trient toutes les étoiles Am et elles seulement,

— si il existe ou non des étoiles Am à raie K normale (la détection des étoiles Am à faible dispersion est-elle donc possible?),

— si Ca peut être normal, ou même un peu surabondant, dans les étoiles Am, i.e. si l'on recherche la cause des étoiles Am, la sous-abondance du Ca est-elle un problème particulier dans les anomalies d'abondance des étoiles Am et dépend-elle de paramètres et de processus physiques propres au calcium?

BIBLIOGRAPHIE

Henry, R. C.: 1969, *Ap. J. Suppl.* **18**, 47.
Henry, R. C. and Hesser, J. E.: 1971, *Astrophys. J. Suppl.* **23**, 421.

DISCUSSION

Rudkjøbing: Greenstein found many years ago that the abundance differences between one metallic-line star and some F stars were a function of the ionization potentials, when determined by a curve-of-growth method (with the same curve). This, to me, clearly indicates that the differences between a metallic-line star and an F star is just that of the atmospheric structure.

The occurrence of metallic-line stars in the Hyades also seems to rule out an abundance-anomaly explanation.

(Reference may be made to a tentative explanation of the phenomenon put forward about 25 years ago in *Annales d'Astrophysique*).

Van 't Veer: We are perfectly conscious of the complexity of the interpretation problems concerning Am stars. But over the last 25 years it has been largely proven by many authors that Am phenomena may be explained by surface abundance anomalies. Reference may be made to the works of these last 15 years: C. van 't Veer, P. S. Conti, F. Praderie, M. Smith. In the last two you can find the new tentative explanation by diffusion processes (cf. also the current work of A. Baglin, S. and G. Vauclair, in *Astron. Astrophys.*).

ULTRAVIOLET SPECTROPHOTOMETRY OF EARLY-TYPE STARS

K. NANDY, G. I. THOMPSON, and C. M. HUMPHRIES

Royal Observatory, Edinburgh, United Kingdom

Abstract. The ultraviolet spectra of B stars obtained with the sky scan telescope in the TD1 satellite have been used to obtain narrow-band magnitudes at several wavelengths. These photometric bands have an effective half-width of 100 Å. We have proposed a two-dimensional classification scheme based on ultraviolet colours, and some preliminary results are presented.

1. Introduction

Existing methods of classification of stars on MK systems are based on line intensities in the visible part of the spectrum. Since the energy distributions of hotter stars reach their Planck maximum in the far ultraviolet, a system of classification based on the observed ultraviolet colours will provide additional information on stellar physical parameters, e.g. effective temperature and surface gravity etc. The sky survey telescope in the TDI satellite gives stellar energy distributions on absolute scale at a resolution of 30 Å and the wavelength coverage is from 1350 Å to 2500 Å. In addition, there is a broad band measurement at 2740 Å. We have already obtained data for several thousands of early type stars, for many of which MK spectral types are not accurately known. In this paper, we shall propose a two-dimensional classification system based on ultraviolet colours.

The basis of classification given here is that after correcting for interstellar reddening the stars can be grouped according to their intrinsic flux distributions, which are determined primarily by their effective temperature (since the spectral types range considered here contains most of the energy of the hot stars). As we have shown in earlier papers (Humphries *et al.*, 1975), luminous stars are increasingly deficient, with increasing $\frac{1}{\lambda}$, as compared to main sequence of same spectral types. Therefore, by proper choice of colours the stars can be separated according to their temperature and surface gravity.

2. Observations and Deductions

In order to achieve greater statistical accuracy, photometric bands of effective widths of 100 Å have been derived from the observed spectra to give magnitudes at several wavelengths including 2500 Å, 2190 Å and 1490 Å. The main photometric error of the ultraviolet magnitudes obtained in this way is $\pm 0\overset{m}{.}04$ for stars brighter than $V = 5\overset{m}{.}0$ rising to $\pm 0\overset{m}{.}12$ for fainter stars. The colour index $(m_{2190} - m_{2500})$ has previously been shown (Humphries *et al.*, 1973) to be a useful parameter for determining the amount of inter-

B. Hauck and P. C. Keenan (eds.), Abundance Effects in Classification, 161–165. *All Rights Reserved.*

stellar reddening without prior knowledge of the MK spectral type. The index changes from -0.5 for B0V to -0.3 for A0V but is very sensitive to interstellar reddening, the colour excess ratio $E_{2190 - 2500}/E(B-V)$ being 2. For a preliminary attempt to separate stars into natural groups having the same ultraviolet energy distributions we have first considered the sample of stars with only a small amount of reddening, as indicated by the observed colour $(m_{2190} - m_{2500})$, assuming that this colour index has a mean value of -0.4 for B stars. This sample is restricted to stars with $E_{2190 - 2500} \leqslant 0.3$ so that error due to uncertainties of interstellar extinction is small.

We have chosen the colour indices $(m_{2740} - V)$ and $(m_{1490} - m_{2740})$ as earlier results indicated that the first primarily determines the colour-temperature while the second is sensitive to both temperature and luminosity. These colours have been corrected for interstellar extinction, using the colour index $(m_{2190} - m_{2500})$ in conjunction with the main extinction law derived from our data (Nandy $et\ al.$, 1975).

3. Results and Discussion

In Figure 1 we present the colour-colour diagram $(m_{1490} - m_{2740})$ vs $(m_{2740} - V)_0$, where V is the visual magnitude. Apart from V magnitudes no other visual data have been used. It is found that points plotted in Figure 1 fall naturally into two groups; most of the observed points lie in a fairly narrow region in the upper part of the diagram (as denoted by crossed circles), and all of these are class III to V. However, a considerable number of points lie significantly lower (open circles), and all of these are found to be of luminosity class I and II. A sudden change in slope in the colour-colour diagram occurs at $(m_{2740} - V) \sim 0.5$. This is caused by the Planck maximum moving longward of 1490 Å for cooler stars.

The separation between main sequence and supergiants is due to ultraviolet flux deficiency of luminous stars. It has been suggested (Humphries $et\ al.$, 1975; Nandy and Schmidt, 1975) that the stars of lower surface gravity are cooler than the corresponding main-sequence stars of the same spectral type. In view of these results, for a two-dimensional classification two extinction-free parameters q_1 and q_2 can be constructed, using $(m_{2190} - m_{2500})$ as a reddening indicator, as follows:

$$q_1 = (m_{2740} - V) - \frac{E_{2740 - V}}{E_{2190 - 2500}} (m_{2190} - m_{2500})$$

$$q_2 = (m_{1490} - m_{2740}) - \frac{E_{1490 - 2740}}{E_{2190 - 2500}} (m_{2190} - m_{2500})$$

The values of $\dfrac{E_{2740 - V}}{E_{2190 - 2500}}$ and $\dfrac{E_{2490 - 2700}}{E_{2190 - 2500}}$ have been determined from the mean

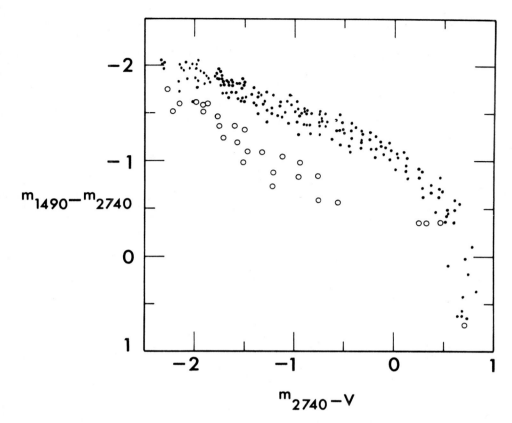

Fig. 1. The plot of $(m_{1490}-m_{2740})$ vs $(m_{2740}-V)$. The stars of luminosity class I and II are shown by open circles; class III to V are indicated by dots.

extinction law derived from a sample of about a hundred reddened stars distributed in different galactic regions (Nandy *et al.*, 1975). For these samples, no significant variation from the mean extinction law has been detected.

Figure 2 shows the plot of q_1 vs q_2 for a further batch of stars which show considerable amounts of reddening as , indicated by the colour index $(m_{2190}-m_{2740})$. Points plotted fall into two groups similar to the distribution as observed in Figure 1. The unique division into groups I and II and III to V is as in Figure 1.

In order to establish the correspondence between the parameters q_1 and q_2, and MK spectral types, mean values of q_1 and q_2 have been determined from all the reddened and unreddened stars studied here; MK spectral types have been taken from the literature (Blanco *et al.*, 1968). These values with the corresponding MK types are shown in Figure 2.

Be stars, however, probably do not fit this pattern. The q_1 and q_2 values of a sample of unreddened Be stars are shown in Figure 3 by circles. It appears that for Be stars of

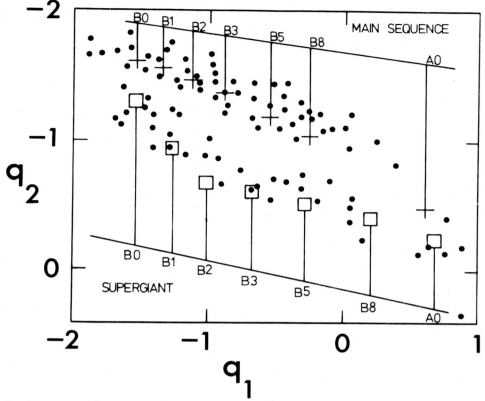

Fig. 2.   The plot of q_1 vs q_2 (see text).

earlier types (B0—B3), q_1 and q_2 coincide with those of supergiants, while for later types they have the same values as main-sequence stars.

References

Humphries, C. M., Nandy, K., and Kontizas, E.: 1975, *Astrophys. J.* **195**, 111.
Humphries, C. M., Nandy, K., and Thompson, G. I.: 1973, *Monthly Notices Roy. Astron. Soc.* **163**, 1.
Nandy, K. and Schmidt, E. G.: 1975, *Astrophys. J.* **198**, 119.
Nandy, K., Thompson, G. I., Jamar, C., Monfils, A., and Wilson, R.: 1975, *Astron. Astrophys.* **44**, 195.

DISCUSSION

Hauck: Have you observed some Be, Bp or Ap stars?
   Nandy: Yes, many of these stars have been observed, but I did not use them in this study since I wanted to stick to objects as normal as possible. Be stars do not fit in the diagram and are located between the class I and V sequences.

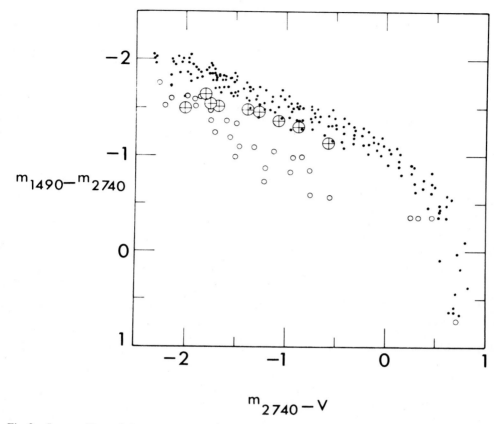

Fig. 3. Same as Figure 1. Be stars are represented by circles with cross.

Houziaux: I would like to mention that the details seen in the spectrum are not just spectral lines. A. Delcroix and myself are analysing these spectra using the spectral synthesis method and it appears that the spectral features result from the convolution of thousands of lines with the wide (30 Å) instrumental profile. The spectrum of B and A stars in the region 1500 to 2500 Å is overcrowded with line absorptions.

Kandel: Since you mentioned the effects of rotation on the structure of stellar atmospheres on the first day, I wonder whether you have any evidence for such effects in your results, especially for B stars. Specifically, do you see any way to distinguish true slow rotators from pole-on fast rotators?

Nandy: We are, of course, interested in this problem, but so far we cannot extract information regarding rotation.

To the second question, the answer is no.

MASS ESTIMATION OF TWELVE K DWARFS

M.-N. PERRIN

Observatoire de Paris-Meudon, France

Abstract. Detailed analyses giving the chemical composition have been carried out for twelve K dwarfs. Iron over hydrogen values have been found. Estimates of the masses of these stars are given.

1. Introduction

Because of the interest of the low-mass stars for galactic structure studies, we began, a few years ago, a study of chemical composition and atmospheric parameters of K dwarfs. Our aim was to correlate results from high-dispersion analyses with spectral classification work.

2. Detailed Analysis

We began this study by reanalyzing nine stars (Perrin *et al.*, 1975), already analyzed by other authors, Oinas (1974) and Strohbach (1970).

In the paper by Perrin *et al.* (1975), were discussed chiefly the strong differences found by Oinas (1974) between the abundance values derived from neutral lines and the abundance values derived from ionized lines of some elements such as Fe, Ti and Cr. The causes of these differences have been found. Indeed, as already said by Dr Bell, we could realize quite perfectly the ionization equilibrium for the nine stars using new oscillation strengths and a multi-branch curve of growth for the Sun (Foy, 1972).

A new paper will soon appear concerning a detailed analysis of three other K dwarfs. Two of these stars, 36 Oph A and 36 Oph B, belong to a visual triple system in which the third component, 36 Oph C (HD 156026), has been analyzed by Strohbach (1970) and by Perrin *et al.* (1975). The other star, HD 191408, is a high-velocity star whose photometry and kinematics have already been discussed by Eggen (1972). Following Eggen, HD 191408 is much older than the stars of the triple system which, according to their strong Ca II H and K emission and their very low space velocities, must belong to a much younger population. These stars, either HD 191408 or 36 Oph A and 36 Oph B, have very well-determined trigonometric parallaxes.

Table I contains some interesting data of the twelve stars studied until now. The heading of each column of this table is self-explanatory. Columns 9–12 contain the results from our detailed analyses.

From this table, we can see that none of the stars have abundance differences with

B. Hauck and P. C. Keenan (eds.), Abundance Effects in Classification, 167–172. *All Rights Reserved*

respect to the Sun, $[Fe/H]^*_\odot$, exceeding the imposed error limit of $\pm$ 0.20 dex; not even the high-velocity star HD 191408.

This is an interesting result and confirms the results by Hearnshaw (1972), da Silva (1975) and Foy (1974) who found other high-velocity stars having normal metal abundances or having a much less metal deficiency than expected.

3. Mass Estimation

The last column of Table I contains the values of the masses which we have attributed to the stars.

These masses have been obtained by using a mean relation between masses of visual dwarf binaries vs $(R-I)$ indices (Figure 1). We found this mean relation using the data from Tables 13 and 15 of a paper by Eggen (1967).

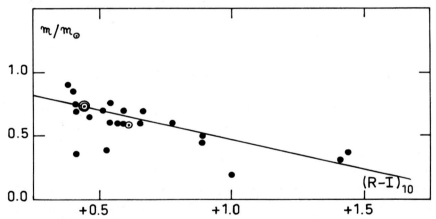

Fig. 1.   Relation between mass and $(R-I)_{10}$ for Eggen's visual binaries. The double open dotted circle stands for 36 Oph A and B, the single open dotted circle for 36 Oph C.

There exists an astrometric mass determination by Brosche (1960) of 36 Oph A and B: $M_A + M_B = 1.46 M_\odot$. Since components A and B are very similar in spectrum and brightness, their masses are probably very close and we could attribute to them the same mass and the same $(R-I)$ index. It is interesting to see that the position of 36 Oph A and B (double open dotted circle) lies on the mean relation in Figure 1.

There exists also a mass estimation by Couteau (1975) for 36 Oph C (single open dotted circle in Figure 1).

The very good fitting for 36 Oph A and B on the mean relation and the rather good fitting for 36 Oph C encouraged us to trust our mass/$(R-I)$ relation and allowed us to estimate the masses of the other nine stars we have already analyzed.

TABLE I

Star	Sp.T.	m_V	$\pi_t \pm$ p.e. (0".001)	$(R-I)_{10}$	M_V	B.C.	M_{bol}	$\log g$	θ_{eff}	$\log T_{eff}$	$[Fe/H]^*_{\odot}$	$\mathcal{M}/\mathcal{M}_{\odot}$
36 Oph A	K0 V	5.1 (1)	189±9	0.44 (2)	6.48	−0.23	6.25	4.6	0.99	3.707	+0.10	0.73
36 Oph B	K1 V								0.99	3.707	−0.02	
36 Oph C	K5 V	6.34 (1)	178±8	0.61 (3)	7.59	−0.58	7.01	4.7	1.11	3.658	+0.06	0.60
HD 191408	K3 V	5.32 (2)	177±8	0.49 (2)	6.55	−0.32	6.23	4.6	1.03	3.689	−0.03	0.71
HD 32147	K3 V	6.21 (2)	109±4	0.49 (2)	6.41	−0.32	6.09	4.5	1.06	3.677	+0.05	0.71
HD 75732	G8 V	5.97 (1)	74±7	0.40 (5)	5.32	−0.16	5.16	4.5	0.97	3.714	+0.22	0.75
HD 145675	K1 V	6.65 (1)	63±7	0.40 (5)	5.70	−0.16	5.54	4.5	0.97	3.714	+0.20	0.75
HD 165341	K0 V	4.24 (4)	195±5	0.38 (6)	5.70	−0.13	5.57	4.5	0.95	3.725	−0.02	0.76
HD 166620	K2 V	6.40 (2)	93±5	0.49 (2)	6.24	−0.32	5.92	4.5	1.01	3.698	−0.02	0.76
HD 190404	K1 V	7.27 (1)	51±11	0.45 (5)	5.80	−0.25	5.55	4.5	1.00	3.703	−0.11	0.72
HD 192310	K0 V	5.73 (2)	120±10	0.45 (2)	6.13	−0.25	5.88	4.5	1.01	3.699	−0.02	0.72
HD 219134	K3 V	5.57 (2)	147±4	0.53 (2)	6.41	−0.41	6.00	4.5	1.05	3.681	+0.04	0.69

Sources: Spectral Type, π, M_V: Woolley *et al.* (1970)

Bolometric corrections: Johnson (1966) after a correction of −0.07

Photometry: (1) Woolley *et al.* (1970); (2) Johnson *et al.* (1966); (3) Gliese (1969); (4) Lutz (1971); (5) Dickow *et al.* (1970); (6) Oinas (1974).

4. Tentative Estimation of He-Abundance

Another contribution of this work on K dwarfs is related to a very important problem.

All the stars have good absolute magnitudes and carefully determined effective temperatures. They have been placed in a (M_{bol}–log T_{eff}) diagram.

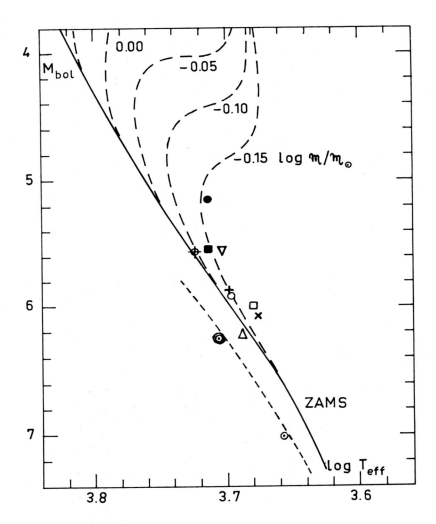

Fig. 2. The (log T_{eff}–M_{bol}) diagram, showing the zero age main sequence (ZAMS) and the evolutionary tracks computed by Hejlesen for $X = 0.70$, $Z = 0.02$ and $l/H_p = 2.0$, together with the positions of the twelve stars studied here, represented by the following symbols: ● HD 75732, ■ 145675, ▽ HD 190404, ⊕ HD 165341, + HD 192310, ○ HD 166620, □ HD 219134, × HD 32147, △ HD 191408, ◎ 36 Oph A and B, ⊙ 36 Oph C; the lower dashed line represents the ZAMS computed by Hejlesen for $X = 0.60$, $Z = 0.02$ and same mixing length parameter.

The positions occupied by the stars have been compared with a theoretical zero-age main sequence (ZAMS) calculated by Hejlesen (1974) for normal solar helium and metal abundances ($X = 0.70$, $Z = 0.02$) and with a mixing-length parameter $1/H_p = 2.0$. In Figure 2, the continuous line represents this ZAMS, the broken line the evolutionary tracks and the different symbols the positions of the observed stars around this ZAMS. For the stars of the triple system the same symbols as in Figure 1 (double and single open dotted circle) have been used. Note that 36 Oph A and B fall together on the diagram because they have the same absolute magnitude and the same effective temperature as derived from detailed analyses.

It can be seen that the stars of the triple system lie below the normal He-abundance ZAMS.

On the contrary, they lie on the ZAMS (lower dashed line on Figure 2) computed by Hejlesen (1974) with higher He-abundance, in respect to the first ZAMS, same normal metal abundance ($X = 0.60$, $Z = 0.02$) and same mixing-length parameter.

We have seen that the metal abundances of the twelve stars are almost the same. If we believe that the position of the triple system on the HR diagram is exact and if we want to justify this position, the only way to do it is to attribute to the triple system a higher He-abundance than that of the other stars in the diagram.

Could this mean that the triple system was formed from interstellar matter enriched in helium? This is an open question and it requires more detailed analyses of low mass visual binaries to be settled.

5. Conclusion

We have seen how carefully determined atmospheric parameters combined with reliable distance parameters can be helpful in the determination of masses and in the estimation of He-content of some K dwarfs.

References

Brosche, P.: 1960, *Astron. Nachr.* **285**, 261.
Couteau, P.: 1975, private communication.
Dickow, P., Gyldenkerne, K., Hansen L., Jacobsen, P. U., Johansen, K. T., Kjaergaard, P., and Olsen, E. H.: 1970, *Astron. Astrophys. Suppl.* **2**, 1.
Eggen, O. J.: 1967, *Ann. Rev. Astron. Astrophys.* **5**, 105.
Eggen, O. J.: 1973, *Astrophys. J.* **182**, 821.
Foy, R.: 1972, *Astron. Astrophys.* **18**, 26.
Gliese, W.: 1969, 'Catalogue of Nearby Stars', *Veröffentl. Astron. Rechen-Inst. Heidelberg*, **22**,
Hearnshaw, J. B.: 1972, *Mem. Roy. Astron. Soc.* **77**, 55.
Hejlesen, P. M.: 1974, private communication.
Johnson, H. L.: 1966, *Ann. Rev. Astron. Astrophys.* **4**, 193.
Johnson, H. L., Mitchell, R. I., Iriarte, B., and Wisniewski, W. Z.: 1966, *Comm. Lunar. Planet. Lab.* **63**.
Lutz, R.: 1971, *Publ. Astron. Soc. Pacific* **83**, 488.
Oinas, V.: 1974, *Astrophys. J. Suppl.* **27**, 391.

Perrin, M. N., Cayrel de Strobel, G., and Cayrel, R.: 1975, *Astron. Astrophys.* **39**,
da Silva, L.: 1975, *Astron. Astrophys.* (in press).
Strohbach, P.: 1970, *Astron. Astrophys.* **6**, 385.
Woolley, R., Epps, E. A., Penston, M. J., and Pocock, S. B.: 1970, *Roy. Obs. Ann.* No 5.

DISCUSSION

Mendoza: Your slide showed a small difference in the trigonometric parallax for 36 Oph A, B than from 36 Oph C. Why not use a single 'mean' value for the three components?

Perrin: Because the determination of parallax for 36 Oph A and 36 Oph B has been made simultaneously whereas 36 Oph C has an independent parallax determination. It is interesting to give this value in the table.

Of course, in the HR diagram I could have given a mean parallax for the three stars, but the position in M_{bol} of 36 Oph C would have been changed by an inappreciable amount.

Maeder: One has to be careful in the comparison of the theoretical zero-age sequence and observations. Firstly, because models with $l/H_p=2$ use mixing-length theory in a case which is known to be not consistent. Secondly, because the opacities used in the quoted models are certainly insufficient for the low mass stars.

Perrin: Of course you are right that Hejlesen's models contain inconsistent l/H_p and poor opacities. But I used ($X=0.70$ $Z=0.02$) for evolutionary tracks calculated with normal He content and ($X=0.60$ $Z=0.02$) for evolutionary tracks calculated for high He content with the same l/H_p and the same opacities. All my stars fall on normal He content tracks except the three very young components of the visual system 36 Oph. These fall on the He-rich tracks.

CAN WE CHECK BY HIGH-DISPERSION ANALYSES STARS BELONGING TO EGGEN'S MOVING GROUPS?

L. E. PASINETTI

Milano-Merate Observatory, Italy

Abstract. We describe the results of analysis of the giant star HD 6497, which belongs to the Arcturus group.

The present discussion concerns a giant star (HD 6497) (Cayrel de Strobel and Pasinetti, 1975), which belongs to the Arcturus group according to Eggen (1974).

This star had been found to be super-metal-rich from narrow-band photometry (Williams, 1971) and metal deficient from the spectroscopic detailed analysis (Cayrel de Strobel, 1966).

We have performed a new detailed analysis employing a very good method proposed by Foy (1971) for reducing the spectrograms. This method allows us to obtain a curve of growth enriched in weak lines. With such a curve (Figure. 1), we have obtained without any doubt a normal iron abundance in respect to the Sun, in agreement with the results obtained from the Geneva photometry (Grenon, 1972), and from Hansen and Kjaergaard

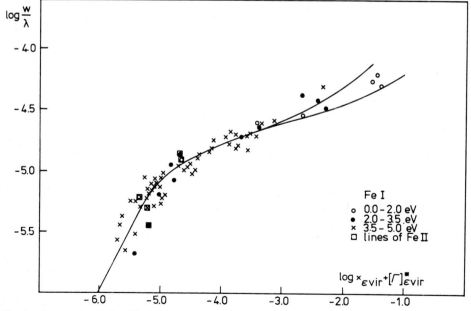

Fig. 1.   Curve of growth of Fe I enriched in weak lines.

B. Hauck and P. C. Keenan (eds.), Abundance Effects in Classification, 173–175. All Rights Reserved.
Copyright © 1976 by the IAU.

(1971). Consequently this new determination has slightly changed the physical parameters which we determine from the neutral and ionized iron lines.

Two remarks are necessary on this result:

– The first remark concerns the spectroscopic techniques: it is easy to understand this
disagreement with the previous spectroscopic analysis, taking into account that in the
former analysis Cayrel de Strobel has not put enough weak lines on the linear part of the
curve of growth. Therefore the possibility exists to draw the curve of growth in two
different ways (Figure 2): small iron deficiency and high microturbulence or normal
metal abundance and low microturbulence.

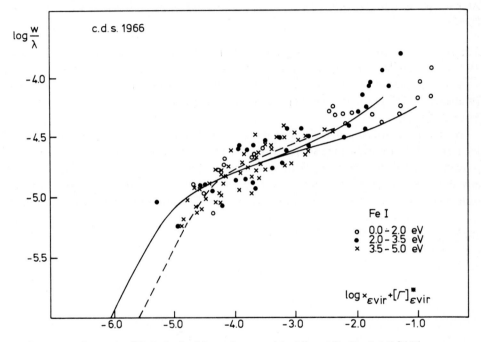

Fig. 2.   Curve of growth of Fe I obtained from the material of Cayrel De Strobel (1966).

– The second remark regards an interesting problem of galactic structure. Eggen had
considered the previous iron abundance i.e. the metal deficiency of this star, as an
indicator of the metal abundance of the Arcturus group. We know (Mäckle et al., 1975)
that Arcturus is surely a metal-deficient star. Therefore several possibilities arise:

Our star may not belong to the Arcturus group or Arcturus itself may not belong to
the group. The group may not be homogeneous in chemical composition or, finally, some
doubt may arise also on the existence of the Arcturus group itself, at least as it has been
given by Eggen.

We have begun at Victoria Observatory an observational programme for testing other
stars of the Arcturus group from a spectroscopic point of view.

References

Cayrel de Strobel, G.: 1966, *Ann. Astrophys.* **29**, 413.
Cayrel de Strobel, G. and Pasinetti, L. E.: 1975, *Astron. Astrophys.* **43**, 127.
Eggen, O. J.: 1974, *Publ. Astron. Soc. Pacific* **86**, 162.
Foy, R.: 1971, *Astron. Astrophys.* **11**, 89.
Grenon, M.: 1972, private communication.
Hansen, L. and Kjaergaard, P.: 1971, *Astron. Astrophys.* **15**, 123.
Mäckle, R., Holweger, H., Griffin, R. and Griffin, R.: 1975, *Astron. Astrophys.* **38**, 239.
Williams, P. M.: 1971, *Monthly Notices Roy. Astron. Soc.* **153**, 171.

DISCUSSION

Williams: Did you use the same spectroscopic plates as in your previous analysis? As to metal abundances of Group stars: I have used narrow-band photometry to investigate abundances of giants assigned to various kinematic groups and apart from finding Hyades (cluster)-like abundances for several stars assigned to the Hyades group, have not found the older groups to be chemically homogeneous.

Pasinetti: No, we have used new spectroscopic plates, but with same dispersion and resolution as in the previous analysis. Regarding the comment: Eggen has given only one abundance value as representative of all the group, precisely the iron abundance of HD 6497.

SPECTRAL CLASSIFICATION OF B AND A STARS FROM DATA OF S2/S68 EXPERIMENT

A. CUCCHIARO

Institut d'Astrophysique, Université de Liège, Belgium

and

M. JASCHEK, C. JASCHEK,

Observatoire de Strasbourg, France

and

D. MACAU-HERCOT*

Institut d'Astrophysique, Université de Liège, Belgium

Abstract. The S2/S68 experiment on the satellite TD1A has supplied, in the wavelength region of 1350 Å to 2550 Å, a very large number of spectra of early stars.

A statistical study, as well as a general analysis of these spectra, has been carried out in order to establish criteria relative to the spectral region envisaged and independent of any previous study in the visible. On the basis of these criteria a system of classification has been outlined.

In a first stage, the spectra of stars visually classed from B0 to A5 have been considered. From four sets of spectrophotometric criteria a two-dimensional ultraviolet classification has been derived.

The purpose of this note is to present a tentative ultraviolet spectral classification of B and A stars from the data of the S2/S68 experiment.

Since the classification in view is a lengthy task, it has been necessary to proceed by successive stages. The first approach was to consider the early B stars. The spectra of the early B stars are characterized by 3 features situated at $\lambda\lambda$ 1400 Å, 1550 Å and 1620 Å and have been attributed mainly to C IV and Fe III (λ 1550) by Peytremann (1975) and Swings *et al.* (1974), Si IV (λ 1400) and Fe II (λ 1620) by Swings *et al.* (1974).

If the following ratios:

$$\frac{R_1}{R_2} = \frac{\text{absolute flux at the bottom of the 1st feature}}{\text{absolute flux at the bottom of the 2nd feature}}$$

and

$$\frac{R_3}{R_2} = \frac{\text{absolute flux at the bottom of the 3rd feature}}{\text{absolute flux at the bottom of the 2nd feature}}$$

are considered and the two values put in a graph, a separation of the sample spectra becomes possible. In fact, it appears that 8 typical regions of the plane can be defined.

Now if the points of region are considered as belonging to one and the same family of points and if for each region a graph, where R_1/R_2 is plotted in function of $r_1 + r_2$, is drawn, three new regions of points can be defined. As an example, the Figures 1a and 1b

---

* Chercheur qualifié au F.N.R.S.

B. Hauck and P. C. Keenan (eds.), Abundance Effects in Classification, 177–180. *All Rights Reserved.*

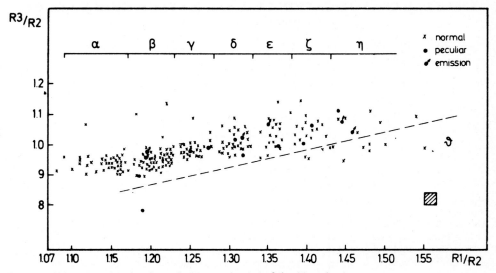

Fig. 1a.   First separation for the early B stars sample in 8 families of points.

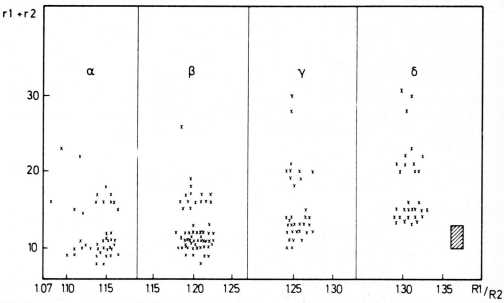

Fig. 1b.   For each α, β, γ, δ families of points, a second separation into three sub-groups.

show the two-dimensional separation for the early B stars. r_1 and r_2 are the intrinsic depth of the 1st and 2nd feature defined from a pseudo continuum (Malaise *et al.*, 1974).

The correlation between this ultraviolet classification and the Yerkes classification both with regard to temperature and to luminosity class gives the first 8 families of points

(α, β, γ, δ, ϵ, ζ, η, θ) corresponding in general to B5, B3, B2.5, B2, B1.5, B1, B0.5, B0 MK type stars.

For each family the three other regions correspond in general to:

— main-sequence stars
— giant stars
— supergiant stars.

The second stage was to consider the late B stars.

A main characteristic of the ultraviolet spectrum of a late B star is the $\lambda\lambda$ 2440 Å feature attributed to Fe II by Swings *et al.* (1974).

If the next quantity:

$$A = 1 - (\frac{F_r}{F_c})_{2400} ,$$

where F_{r2400} is the absolute flux at the bottom of the feature and F_{c2400} is the absolute flux calculated for the pseudo continuum at 2400 Å, is plotted as a function of $m_{1400} - m_v$, a separation of the sample into four families of points is also possible.

Now, as for the early B stars, if $m_{1400} - m_{2730}$ is given vs the A value for each family of points, a second subdivision appears.

Correlation with the MK spectral types shows that the first 4 families correspond to B6, B7, B8, B9 and B9.5 stars.

The sub-groups for each family correspond to
— main-sequence stars
— giant stars
— supergiant stars.

The last step was to analyze the early A stars. As for the late B stars, a main charac-teristic of the A ultraviolet spectrum is the $\lambda\lambda$ 2400 Å feature but attributed for the early A stars to Ni II and Fe II by Gros *et al.* (1973).

If the following ratios:

$$\frac{F_{2400}}{F_{2300}}$$

and

$$\frac{F_{2400}}{F_{2730}}$$

are taken into account, one can show that, as for the two preceding stages, five regions of the plane can be well defined and the correlation with the MK spectral type shows that the ν, ξ, o, π and ρ families correspond in general to A0, A1, A2, A3 and A5 stars.

Conclusion

The preceding considerations show that the ultraviolet spectra of B and early A stars can be classified my means of ultraviolet criteria exclusively.

For more details a substantial paper will be published later.

References

Gros, M., Sacotte, D., Praderie, F., and Bonnet, R. M.: 1973, *Astron. Astrophys.* **27**, 167.
Malaise, D., Beeckmans, F., and Jamar, C.: 1974, *European Regional Meeting in Astronomy*, Trieste, (in press).
Peytremann, E.: 1975, *Astron. Astrophys.* (in press).
Swings, J. P., Jamar, C., and Vreux, J. M.: 1973, *Astron. Astrophys.* **29**, 207.

DISCUSSION

Gerbaldi: Did you observe some Am stars and what did you notice?

Cucchiaro: Yes, we have observed Am stars; It appears that some of them escaped from our classification scheme and others agreed with it. But the number of Am stars of our sample is very low until now and any conclusions can be outlined on an eventual separation between Am and normal A stars.

McCarthy: What do you observe as peculiar features in the location on your plots of the Be stars as compared to the non-emission B stars?

Cucchiaro: The Be stars as compared to the non-emission B stars don't present striking characteristics in this wavelength region. Their place in the present classification, except for a few of them, corresponds to what is known from visual spectroscopy.

Garrison: Perhaps I missed an important point, but I don't understand why the results shown in your diagrams are quantized according to MK types and luminosity classes. Is it that God believes in MK classification?

Cucchiaro: We have asked ourselves such a question and until now no satisfactory answer has been found. Perhaps the number of spectra is too small, yet I believe that to give a real answer we can analyse all the physical processes which correspond to our criteria.

PART V

ABUNDANCES IN STELLAR POPULATIONS

ABUNDANCES IN STELLAR POPULATIONS

H. SPINRAD

Department of Astronomy, University of California, Berkeley, Calif., U.S.A.

Abstract. Stellar abundances are reviewed with emphasis on large-scale effects which may yield clues to galactic structure and evolution. Spectroscopic and indirect photoelectric abundance criteria are discussed, and utilized.

The abundance statistics of nearby galactic disk stars, dominated by M dwarfs, but observed at spectral types F and GV and K III, suggest a weak age-abundance relationship with a substantial dispersion at any time. Very metal-poor stars are extremely rare. Spatial abundance gradients, with higher metal abundances occurring nearer the galactic centre, are indicated. Disk abundance gradients are prevalent for light elements in other Sb and Sc galaxies.

The confusing status of supermetallicity is again reviewed. The super-metal-rich (SMR) *giants* (like μ Leo) are either over-abundant because of self-N-enrichment (from C−N−O processing?)and boundary-temperature cooling, or are really SMR. Each case may be reasonably argued. The old galactic clusters M67 and NGC 183 seem, by recent *indirect* acclaim, to be only slightly more metal-rich than the Sun. The Spinrad-Taylor data on the M67 giants would still seem to superficially suggest over-abundances in Na and Mg, but other interpretations are possible.

SMR dwarfs, like HR 72, and subgiants, like 31 Aql are surely very old, and have metal abundances larger than the Hyades. However, they are, by number, only $\approx 5\%$ of the local main sequence.

The galactic halo star tracers − red giants and RR Lyrae stars, have been observed extensively, lately. There is some indication of an abundance gradient from 5 or 10 kpc galactocentric radius out to $r \sim 100$ kpc. The most metal-poor stars observed in the Draco system are about 1000 times less abundant in heavy elements than is the Sun, and much of the galactic disk.

Abundances in other galaxies, as a function of their total mass, and stellar/gaseous composition are also reviewed. There is a clear dependence of abundance on galaxian total mass.

1. Introduction and Emphasis

This review will emphasize stellar abundance determinations on a large scale − as applied to the populations, which can yield large-scale information on the structure and past evolution of our Galaxy, and other galaxies.

Also since we deal with galactic archeology, I will almost ignore the problems of young stars and spectral types O, B, and A in this talk. All this means that we have to observe and also understand the spectra of many faint (distant) stars of spectral type F5 and later, or the integrated spectra of millions of stars. Then right away we have a strong operational constraint on the information content of the data and the depth of analysis to be expected. We have to look at indirect, coarse criteria for abundances − not the conventional, high-dispersion spectroscopic analysis. But large, homogeneous series of photoelectric observations of metallicity criteria (see the review by Bell in this Symposium) have advantages, too. As Williams (1974) states, the narrow-band technique has the advantage of speed, homogeneity and impersonal comparison. Of course, the detailed analysis of some interesting stars, difficult as they may be, is very necessary.

B. Hauck and P. C. Keenan (eds.), Abundance Effects in Classification, 183−204. *All Rights Reserved.*
Copyright © 1976 *by the IAU.*

Now what are these indirect or semi-direct criteria we can use? Most closely aligned to conventional spectroscopic analysis are photoelectric measures of individual strong lines or groups of them – pioneered by the Cambridge group under Redman and Griffin, and recently by P. M. Williams (1971a, b). Then we have groups of weak lines, isolated by an echelle spectrograph, observed by Gustafsson *et al.* (1974); then the medium-resolution scans of Spinrad and Taylor (1969), and finally many other types of intermediate-band photometry – such as Strömgren photometry, DDO photometry which measures CN [$C(41-42)$], and recently a system extending to the near-IR developed by R. Canterna at the University of Washington. Even *UBV* photometry, or *RGU* photometry can be applied to stellar abundance determinations, perhaps in restricted parts of the HR diagram or together with other data. Naturally, we find the very narrow-band or single--line techniques applicable to mainly bright stars, and in the context of this talk, that usually means nearby K giants or a few very nearby G dwarfs. The broader-band, (often) less discriminating tests, can be applied to luminous stars almost across the galaxy, and to lots of main-sequence stars. What is desired is the best of the two worlds – and perhaps the next generation of digital, sky-subtracting scanners and panoramic detectors will provide the means. A start has been made by Butler and Kraft (1975).

Until then, we use what we can. An example of what can be done with the blanketing measures of, *U–B, B–V* colours and the unique temperature dependence of *(R–I)* or *T* (Spinrad and Taylor, 1969) follows:

If we examine the *U–B, B–V* plane positions of nearby G-K dwarfs and K giants, vs the Hyades cluster sequence of Johnson and Knuckles (1955) and Johnson, Mitchell and Iriarte (1962), we note a large scatter, and rather poor separation of later-type stars of differing metal content. However, if we can group either dwarf or giant stars of nearly identical red colour (and therefore, effective temperature) together, we note that their *loci* in the *U–B, B–V* plane are short lines, whose slope $S = \dfrac{\Delta(U-B)}{\Delta(B-V)}$, and is a function of *T* or *R–I*. Table I lists the results, seen in Figure 1. I call these slopes empirical blanketing vectors. In practice this is not very different from the work of Becker and Steinlin (1956).

The usefulness here is that abundances could probably be interpolated for stars with *B–V* $\geqslant 0.7$, with only a luminosity class, *UBV*, and *T*, or *R–I* known. This is quite practical for stars to *V* $\simeq 15$, and thus a large fraction of the galaxy is available *in situ* for K0-K3 III types.

2. Abundance Results on Stars near the Sun – The Galactic Disk

Pagel and Patchett (1975) have recently compiled abundance data from several statistically complete samples of nearby dwarf stars.

They find for the long-lived G dwarfs the abundances, relative to the metal-rich Hyades cluster, were log-normal in [Fe/H], with a mean at $\simeq-0.3$. The age effect is

TABLE I

Empirical blanketing vectors for K stars

a) *K dwarfs:*

T	$(R-I)$	Equiv. Sp. Type	$S = \dfrac{\Delta\,(U-B)}{\Delta\,(B-V)}$
347	0^m36	G8V	4.0
376	0^m40	K0V	4.4
442	0^m53	K3V	6.7
485	0^m55	K4V	8.5
540	0^m66	K5V	10:

b) *K giant stars:*

392	0^m47	G9III	3.0
445	0^m57	K2III	4.5
500	0^m70:	K3-4III	6.6

present — a modest, but widely dispersed heavy element enrichment $Z(t)$ — with considerable scatter among field stars of a given age. The data are noisy, but according to Pagel and Patchett, there is a significant increase of Z with t — with no early peak. Note that the lowest mean Z considered is only at about 1/3 the *solar* metal abundance at an age of $l \simeq 10$ x 10^9 yr. Thus there is evidence, long known for clusters, that the metal-enrichment galaxy from some very low primordial level, was nearly half-complete when the disk stars formed. However, it has slowly increased since that time, if the Pagel and Patchett statistics are accurate and appropriate.

It is a pity that, at present, we cannot do more — on a quantitative level — with abundances for the M dwarfs. They are the vast, silent majority of stars in the solar neighbourhood and the nearby Universe. Scanner abundance determinations, such as those by Spinrad (1973) are a start — they show some strong — and some weak-line stars of both high and low velocity. It is difficult to use the scans alone for the badly needed quantitative analysis of M dwarfs. However, Barnard's star and Kapteyn's star, both high velocity M dwafs, have very different line strengths — presumably Barnard's is rich and Kapteyn's is metal-poor.

The oldest disk stars, as signified by their locations slightly above the main sequence (or as subgiants) have been observed spectroscopically by Hearnshaw (1972, 1973, 1974 and 1975). Other old stars (G and KV) can be found on the basis of abnormally weak chromospheric reversals at the centres of H and K; quantitative analysis by M. Penston shows many nearby stars to be older than the Sun, as expected.

The *oldest* disk stars have a rather large metallicity spread — from [Fe/H] $\simeq -0.8$ to even +0.44. Some of them are apparently SMR; we talk about this later.

Hearnshaw (1975) has suggested that old, higher velocity disk stars have higher [C/Fe] ratios than the average of disk stars with the same (low) iron abundance, but all the *Fe-rich* stars are carbon *overabundant*. The work of Oinas has suggested [Na/Fe] $\approx$

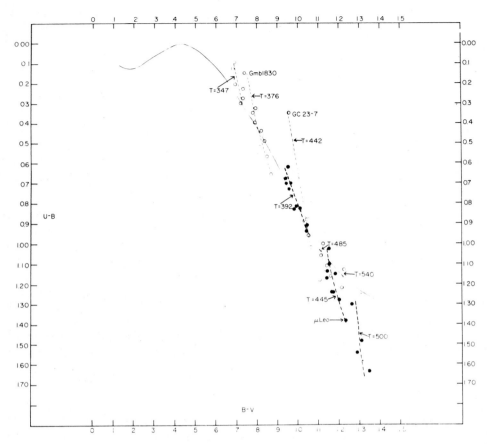

Fig. 1. Empirical blanketing vectors for K giants and late G-K dwarf stars. Points plotted and connected are stars with very similar red colour, *T*. The solid line in the Hyades main-sequence. Note the steepening of the vectors for the cooler stars.

[Fe/H]. Here we could begin to assemble clues important to early disk nucleosynthesis in the galaxy. Again the practical need to observe many faint stars suggests relative measures of CH (the G-band), Fe and T_{eff} by a (new) standard photoelectric system. That should be straightforward.

The statistics of normal disk K giants can now be studied with the large amount of data available on their DDO indices — mainly by Janes and McClure. In their 1971 and 1973 papers, Janes and McClure studied the kinematics of 799 K giants with photometry; the cyanogen index, their metallicity parameter, is rather well correlated with spectroscopic [Fe/H], except in some (unusual?)stars with N enrichment. Studying kinematics of their sample, they found that the K giants in the *weakest* CN group have a predominance of negative θ velocities — i.e. they are in rather eccentric orbits, and are likely old stars. However, the *strongest* CN stars also have predominantly negative θ velocities. These stars, with δCN $> 0\overset{m}{.}06$ (above the Hyades giants in CN) are 10% of the local K giants, in Janes (1972) thesis. Normal (δCN ≈ 0) giants cluster near the LSR in the Bottlinger diagram. However, in the Z velocity domain the situation is more conventional; the dispersion in Z velocities is a function of the CN anomaly, in the sense that the very weak CN stars have relatively high Z velocities, while the strong CN stars do not — they stay near the galactic plane. Thus the negative θ velocities (lag in rotation) of the strong-CN stars must be due to a *radial gradient* of CN strength in the *plane* of the galaxy. If these local gradients are blindly extrapolated to the centre of the galaxy, the population there — near the plane, would be dominated by the strong CN stars while in a larger volume around the nucleus the very weak CN stars would be very important. However, our galaxy must have a bulge population too, which we hardly sample out here, so this extrapolation in 10 kpc is too great. The Janes and McClure data yield a radial gradient in '[Fe/H]' of +0.023/kpc. Thus if $R_0 = 9$ kpc, the nuclear (*disk*) value of the CN-derived [Fe/H] would be +0.2, a modest overabundance. Data on other galaxies, mainly from H II region emission analysis, suggest that a linear extrapolation (at least for elemens O, N...) is too modest — I'll talk about that later.

The main thrust of this argument is to show that the heavy element abundance is a function of position of birth in the (exponential) disk for disk stars — now relatively near the sun. This conclusion is strenghtened by the McClure *et al.* (1974) observation of the medium-age *galactic* cluster, NGC 2420 — toward the anti-centre. It has a low metal abundance, δCN $= -0.06$ which corresponds to [Fe/H] $= -0.4$, the lowest of any well-observed galactic cluster! Studies of other old to moderate age clusters, in the Perseus arm and beyond, are clearly desired.

3. The Status of Supermetallicity

I'd like to discuss the status and observed number of super-metal-rich (SMR), or apparently SMR stars, in 3 somewhat controversial or at least confused situations: They are:
1) The field K giants,
2) The old galactic clusters M67 and NGC 188,
3) The unevolved main sequence, F5-K7V.

In each section there are certain contradictions which I would like to mention — but cannot necessarily clarify!

It is surprising to this reviewer to find different rather sophisticated observations and analysis still yielding self-contradicting results for SMR K giants, especially the Spinrad-Taylor prototype star, μ Leonis (K2 III).

SMR stars are moderately rare, but a non-negligible fraction of nearby stars. Table II lists the proportions. Let's discuss recent work on field SMR K giants. There have been three major efforts to check on the abundances of prototype giants by conventional means. Blanc-Vaziaga *et al.* (1973) have analyzed ϕ Aur and μ Leo vs ϵ Vir, using high-dispersion spectra and theoretical line strength computations, based upon a grid of model

TABLE II

Percentages of stars classified 'SMR'[*]

1) *K giants:*		
Author(s)	% SMR	Technique or Parameters
Janes (1972)	10%	DDO CN index.
Spinrad and Taylor (1969)	~10%	Scanner line indices
Gustafsson *et al.* (1974)	~ 6%	Echelle spectrometry of metal lines.
Average =	9%	
2) *F-K dwarfs:*		
Taylor (1970)	4%	G-K5V scanner lines.
Raff (1975)	~5%	F5-G4V high-disp. spectra and models.
Average =	4,5%	

[*] Indirect or direct abundance ≥ Hyades stars.

atmospheres. More recently, Ruth Peterson (1975) has obtained and analyzed high resolution spectra of μ Leo and two less-strong-line K2 III stars, κ Oph and ι Dra. These stars have very similar red colours, and presumably the same T_e. Also, V. Oinas (1974) has utilized high-resolution coudé spectrograms of SMR giants (especially HR 8924, α Ser, μ Leo, HD 112127, 18 Lib A and θ UMi) and 'Atlas' models to compare red giants to the Hyades and the sun, and then to derive detailed abundances.

All three investigators need three basic and simple assurances to do their job:
(a) That the atmospheric structures of these giants are comparable,
(b) That the relative temperatures can be accurately determined (a negligible problem for Peterson), and
(c) That the equivalent width measurements are self-consistent among all compared stars.

It would seem that item (c) *should* be satisfactory, and I have not searched-out cross checks; (b) is crucial and everyone knows it or minimizes it by observing stars at a fixed red colour, and the first question (a) is assumed or anticipated, or tested later. Table III tabulates some recent spectroscopic and indirect results on SMR star abundances.

Blanc-Vaziaga, Cayrel and Cayrel's analysis suggests φ Aur is normal in Fe group elements, mildly overabundant in Na, while μ Leo is overabundant in Na and Ca, and normal in Fe and most other metals. They suggest that 'supermetallicity' is highly selective in the periodic table!

Oinas' technique is, with slight exceptions, very similar to the above; he used model atmospheres to compare the giants to the Sun. There may be some problem remaining, as Oinas found he had to alter the derived spectroscopic gravity to make abundances from ions and neutral coincident, and surprisingly, he found the most extreme K2-K3 III SMR stars to be modestly overabundant in Fe, Cr, Ti and V (averaging +0.2 dex), but up by 0.5 in [Na/H]. So at least Oinas found a substantial Na excess, in accord with Blanc-Vaziaga *et al.* But the basic conclusions differ for Fe and Cr; the situation (as I see it) is not clarified

TABLE III

Recent spectroscopic and photometric metallicities for SMR stars

a) *Giants:*

Author(s)	Technique	Typical or maximum $[Fe/H]_\odot$	Stars involved
R. Griffin (1975)	Coudé spectra	−0.3	M 67, IV-202 Red Giant
R. Peterson (1975)	Coudé spectra	+0.0	μ Leo compared to ι Dra and κ Oph
P. Williams (1974)	Coudé spectra	+0.3	ι Dra
V. Oinas (1974)	Coudé spectra	+0.1, +0.2 (Na up)	μ Leo other SMR KIII
Blanc-Vaziaga *et. al.* (1973)	Coudé spectra	−0.10 (Na up)	μ Leo, ϕ Aur.
R. Canterna (1975)	Broad-Band Red-Green photom.	+0.3	Strong-CN stars, μ Leo, 20 Cyg
Gustafsson *et. al.* (1975)	Echelle Photo-electric spectra	$\sim$+0.3	Strong-CN stars, μ Leo, 20 Cyg

b) *Dwarfs and some slightly evolved Sub-giants:*

M. Raff (1975)	Coudé spectra	a few up to +0.4	η Boo, F8-G2V
J. Hearnshaw (1974, 1975)	Coudé spectra	5 with $\geqslant 0.3$	G V and GIV
V. Oinas (1974)	Coudé spectra	+0.25, (Na up more)	K dwarfs, ρ, Cnc, Hr 1614
M. Grenon (1973)	Geneva Photometry	to +0.45	G-K V

by Peterson's result; μ Leo, κ Oph and ι Dra are all found to be at the solar level in [Fe/H]! She found that the weak ($w_\lambda < 100$ mÅ) lines for all three K2 III stars were about the same, and thus, since she found κ Oph had [Fe/H] = 0.0, so did μ Leo and κ Oph. However, she did find a strengthening of all lines formed near the boundary layer and suggests (as did Strom *et al.*, 1971) a lowered boundary temperature for the highly blanketed μ Leo. This is plausible, since μ Leo has very strong bands of CN. Thus Peterson concludes that μ Leo *appears* 'SMR' because of a low boundary temperature, which strengthens low-excitation lines of neutral atoms! This is a smooth scenario, but a likely flaw in the analysis is the choice of comparison stars, as both Spinrad-Taylor and Janes find that κ Oph has somewhat atypically strong blanketing in $U-B$, stronger-than-normal Ca I, Mg I and (slightly) Na I. Its DDO CN anomaly is +0.041, 2/3 the way to the Hyades! Moreover, Williams (1974) found ι Dra to be metal-rich by a factor of two, compared to ϵ Vir (usual ref. giant); the inconsistency in this result, the conclusion of Blanc-Vaziaga *et al.* and Peterson is sharpened when we note that Peterson finds the weak ($W_\lambda < 100$ mÅ) lines in the spectrum of ι Dra to be only 2% stronger than those in κ Oph, the standard of normalcy in her selection. Someone has to be incorrect!

Just as serious, in my thinking, is the difficulty of relating the relatively large narrow-band-indices for SMR giants to their detailed abundances. Do the correlations between spectroscopic [Fe/H] with an index, like CN or Canterna's $M-T_1$, break down only at [Fe/H]$_\odot \gtrsim 0.2$? That would seem logical only if boundary cooling due to strong CN took over. However the high-resolution echelle spectrophotometry of Gustafsson et al. (1974) would seem to rely on lines near or strictly on the linear part of the curve-of-growth. These Fe lines are insensitive to boundary temperature fluctuations. These authors measured 80 G and K giants and utilized scaled solar model atmospheres, and found the Hyades to have [Fe/H]$_\odot$ = +0.16 (good agreement with previous conventional analysis). Some giants, including μ Leo, were found to be SMR. These procedures seem sound, too. I cannot reconcile their result with much of the conventional spectroscopy, of which we are familiar. Both seem to be credible.

Other narrow-band measures, such as the stronger Fe lines observed by Williams (1974), suggest SMR stars to be overabundant — as did Spinrad and Taylor for Na, Ca, and Mg.

One last remark on the SMR K giant problem should be re-stated; we probably cannot conclusively decide on how *superficial* 'SMR-ness' may be. It could represent a modest increase in Na and N and nothing else at all in an evolved star — or it could mean a primordial increase in abundance for almost all the elements. But we know quite empirically, that strong CN stars are more numerous toward the galactic centre (Janes and McClure 1973), so something physical is happening! If Grenon (1973) is correct and the same thing happens for G dwarfs and subgiants, we can be sure it's more basic and important than the CNO-generated N^{14} increase! On the other hand, the proportions of nearby SMR giants is twice that of the dwarf SMR sample (Table II), so that one could argue that half the SMR giants are, indeed, caused by self-enrichment in N^{14}, which makes strong CN. Then this yields a positive CN residual or overblankets the atmosphere of the K giant. Another counter argument: Dearborn et al. (1975) find μ Leo has even less C^{13} (relative to C^{12}) than does α Boo — μ Leo has a normal giant C^{12}/C^{13}, despite the strong CN bands, which imply much N^{14}. So the selfmixing arguments are not complete, for at least μ Leo.

The status of nearby SMR field *dwarfs* and slightly evolved subgiants is, for a change, almost unanimously in favour of finding a small percentage really SMR. Taylor and Raff each suggest $\approx 5\%$ to be SMR. The most blatant examples are 31 Aql (Hearnshaw, 1972), HR 511 and HR 7670, (Hearnshaw, 1974), HR 72 (Spinrad and Luebke, 1970), ρ' Cnc, 14 Her, Hr 1614 (Taylor, 1970; Oinas, 1974), δ Pavonis (Harmer et al., 1970) and η Boo (Raff, 1975).

I illustrate the spectrum of HR 72 and its Fe I curve-of-growth in Figures 2 and 3; the line strengths are really impressive. These are basically old to very-old stars, from the H-R diagram position of the subgiants, and the lack of measurable H and K reversals in the G-K dwarfs.

The actual abundances seem safe, because the stars are usually compared to the sun, and the comparisons often involve a small difference in temperature. Also we think that the atmospheres of solar-like stars are simpler or, anyway, more so than those of giants. Boundary temperature lowering by molecular blocking should be minimal, at least down to type K2 V.

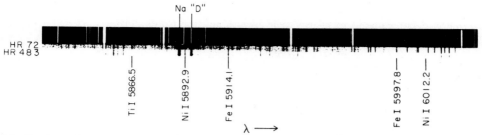

Fig. 2. A portion of the red spectra of HR72 (SMR G star) and HR483, a slightly metal-rich G2V. Note the strength of the weak lines and the strong, broad wings of Na 'D' in HR72. The original dispersion was 8 Å mm⁻¹.

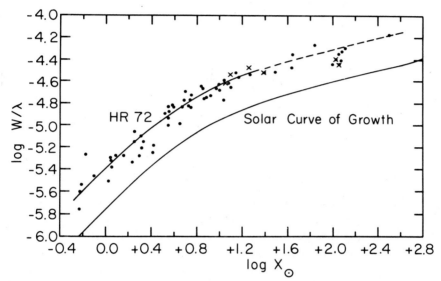

Fig. 3. The resulting curve-of-growth for HR72, compared to that for the Sun. From Spinrad and Luebke (1970). The overabundance of Fe is a factor of 2.5 in HR72.

Still, there are some surprises: Oinas still finds a Na-excess in the SMR dwarfs; [Na/Fe] ≃ +0.2. Taylor's (1970) survey finds more K V stars with Na 'D' > (Na 'D')$_{Hyades}$, than stars with Fe I excesses. This might be an interesting lead to follow up.

Now the story of the old galactic clusters and supermetallicity: Spinrad and Taylor (ST) (1969) had suggested SMR line strengths for both M 67 and NGC 188, the oldest well-observed galactic clusters. This conclusion has been doubted and criticized ever since, and, I am sorry to say, has not been supported by any new evidence at all. I do *not* believe the ST scans of the M 67 giants have a systematic error, but I can't convince anyone. The old clusters are important — because of their age — and accurate abundances are needed.

The most direct new evidence on M 67 is the low-dispersion analysis of F and G dwarfs by Barry and Cromwell (1974) and the high-resolution spectroscopic work by R. Griffin

(1975) on the cool giant, M 67, IV -- 202. Barry and Cromwell find M 67 only slight-ly metal-rich compared to nearby field dwarfs and the Coma cluster stars, while Griffin suggests the M 67 giant to be under-abundant by about a factor of 2 with respect to the Sun. Barry and Cromwell's H measuring technique can be used to derive the reddening to M 67, and they got rather lower reddening than do others. If we demand $E(B-V) = 0^{m}\!\!.06$ (Eggen and Sandage, 1964) or $E = 0^{m}\!\!.07$ (from HI in that direction — perhaps a risky cor-relation), then the abundances will go up somewhat, perhaps to $[Fe/H]_{\odot} \simeq +0.1$ or $+0.2$, but probably not over the Hyades. The Griffin analysis seems straightforward enough, provided the relative stellar temperatures are well determined, and the abundance of α Boo is known. However the metal-deficiency of M 67 makes little sense if we examine the various cluster CN anomalies assembled by Boyle and McClure (1975). We tabulate in Table IV these results, in order of $\delta(CN)$.

TABLE IV

δ CN for Stars, Galactic and
Globular Clusters

Star or Cluster	$\delta(CN)$
μ Leo	$+0^{m}\!\!.11$
Praesepe-Hyades	$+0^{m}\!\!.07$
M67	$+0^{m}\!\!.03$
NGC 188	$+0^{m}\!\!.00$
[Field K giants]	$+0^{m}\!\!.00$
47 Tuc	$-0^{m}\!\!.07$

And also Pagel (1974) finds M 67 and NGC 188 to have CN indicative of metal abun-dances between the Hyades (+0.2) and the Sun (0.0).

Unless the average field giant is very metal-poor, then M 67 has to be intermediate on these CN systems. Of course a direct measurement of Fe would be better.

Probably what is really needed now, is a photoelectric measurement of medium-strength Fe I lines in M 67 stars below the turnoff, near G2V — where both the Sun or the Hyades can be used as controls.

In any case, a slightly-above-solar-abundance level for the old clusters is, perhaps, still a little uncomfortable for the gradual disk enrichment picture outlined earlier.

4. The Galactic Halo — Abundances from Tracers (and Occasional Field Samples)

It's been acknowledged that the halo population is quite metal-poor for some time (Lindblad 1922, Schwarzschild and Schwarzschild 1950). A considerable amount of qualitative information about [Fe/H] exists from the work of Mayall (1946), Morgan (1956) and Kinman (1959). Red giants in globular clusters were studied quantitatively by Helfer *et al.* (1959) and their very metal-poor field analogues were observed by Wallerstein *et al.* (1963).

There is a definite correlation between [Fe/H] and space motion for nearby halo stars, *selected* on the basis of proper-motions. These very high velocity stars are rare, per unit volume($\sim$1% near us); however by the 1960s enough of them had been found, for which photometry was available, to allow Eggen, Lynden-Bell and Sandage (ELS) (1962) to correlate stellar ultraviolet excesses with space motions, especially the |W| (the Z-component). This led to the ELS picture of a rapid halo collapse with most nucleogenesis complete during the dynamically brief ($\sim10^8$ yr) collapse. The classic ELS paper is still important; it has set the stage for many important subsequent investigations and has carried over a few prejudices, too.

Now we have the technical capability to study a few nearby high velocity stars of the halo population in great detail (cf. Sneden, 1973 and Pagel, 1972), and many distant 'tracers' — RR Lyrae variables and globular cluster stars — in a coarse, comparative way. Despite the large distances involved, the study of luminous, evolved stars should allow us, in time, to map out the present distribution of metals in the galactic halo (presumably frozen in since the early days), and then evolutionary modelling should be successful.

The field halo star work I'd like to mention is a study by McClure (1973), complimenting the earlier work of Sturch and Helfer (1972). McClure noted that the CN anomalies of field K giants at the North Galactic Pole became more pronounced and negative (metal-poor) at heights of 1 kpc and more above the plane — but the dispersion in δ CN was still large at this distance — occasional strong-CN stars are still present in the near halo, although at $Z \geqslant 2$ kpc the few stars sampled seemed to be as metal-poor as Arcturus, judging from their CN bands. There are no extremely metal-deficient stars in this sample, or the one observed by Sturch and Helfer. Thus the halo stars selected by position *in situ* are moderately metal-poor — surprising no one. However, we recall that the *average metal-poor* star, selected not on the basis of motion or position far from the galactic plane, is in an almost circular orbit (Bond 1970)!

Unfortunately, to go further we have to pick on special tracer stars — variables in the field, in clusters, and globular cluster red giant stars. Some of the most recent work on these objects is by Butler (1975), Butler and Kraft (1975) and by Kraft (I thank Bob Kraft for many discussions and permission to quote some of his results of work in progress...).

The metal abundances of RR Lyrae stars, in the field and in clusters, were first studied by Preston (1959, 1961). This was done by estimating equivalent spectral types for both the H Balmer lines and the Ca II K-line. Preston's ΔS parameter, defined as $\Delta S = 10$ [Sp(H)–Sp(K)] correlated with the periods of the nearby RR Lyrae type 'a' variables, and their space motions. But because of the long exposure times demanded by a photographic spectral survey of distant RR Lyrae, this ΔS system hasn't enjoyed the popularity associated with intermediate-band photometric systems. The last few years Butler, and Butler and Kraft have used the new image-tube-scanner (ITS) (Robinson and Wampler 1972) at the Lick 120″ reflector to obtain low dispersion digital spectra of both field and cluster RR Lyrae stars. Under average observing conditions it takes 15^m to obtain a good ITS spectrum (in the blue) at $B\sim$ 15. We illustrate in Figure 4 the spectra of some cluster RR Lyrae and 2 Coma main-sequence stars; note the relative weakness of.

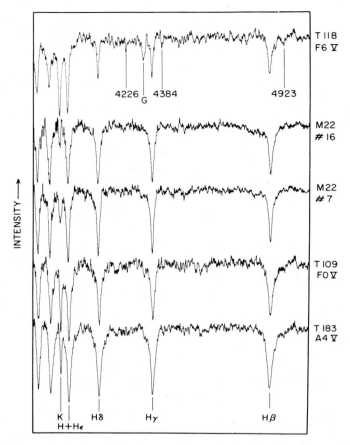

Fig. 4.   ITS spectra of selected RR Lyrae stars in M22 illustrated with some Coma cluster A and F star standards. Note the weakness of Ca II K and the G-band in the metal-poor cluster variables. From Butler (1975).

Ca II in the M22 variables. Figure 5 shows Butler's plot of the equivalent widths of H Balmer lines and the K line, vs Spectral Type. The ΔS for distant globular clusters were corrected for the possible interstellar K-line components, and a calibration of the new photoelectric ΔS index was established by Butler from analysis of high-dispersion spectra of the brightest field RR Lyrae stars. The calibration was:

$$[Fe/H] = -0.16 \ \Delta S -0.23 \qquad\qquad (1)$$

The resulting abundances from ΔS correlate well with Sturch's (1966) $\delta \ (U-B)$ values and Jones (1971, 1973) K-line index. The most metal-poor field stars ($\Delta S \sim 12$) have [Fe/H] $= -2.1$ while $\Delta S = -1$ applies to a star of nearly solar composition – and such field RR Lyrae do exist, although they haven't been located in any globular cluster.

The application of this technique to mapping out RR Lyrae abundances deep in the

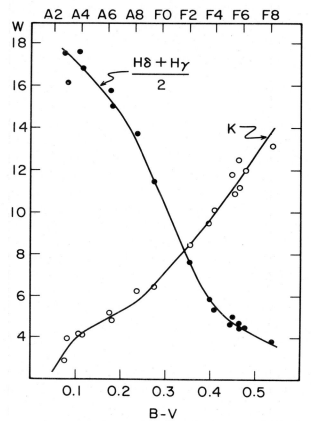

Fig. 5. Butler's plot of the equivalent widths of the H-Balmer lines and the Ca II K line, as a function of Spectral type, for normal stars. Such stars would have $\Delta S \simeq 0$.

halo is obvious; results so far on globular clusters compare well with Canterna's red-green photometry (1975) of the $(M-T_1)$ index. Table 5 lists abundances of globular clusters from Canterna's photometry of the giants (independent of C, N, O), Butler's [Fe/H] from RR Lyrae ΔS values, and the DDO CN photometry of Hartwick and McClure (1974) and Osborn (1973).

The agreement is fairly good. Under present investigation by Kraft are RR Lyrae near the galactic centre — in the field of NGC 6522. The period-frequency distribution of stars there (Hartwick, *et al.*, 1972) seems to mimic that of M5, for which [Fe/H] ~ -1. But the giants near the galactic centre, according to both Arp (1965) and van den Bergh (1972) are likely metal rich — as M67 or NGC 188. Do the RR Lyraes and giants of the galaxies nuclear bulge belong to a different population, or are there selection criteria — e.g. — *no large pulsations* in horizontal-branch stars if [Fe/H] $\gtrsim 0.0$? Anyway, Kraft's preliminary data show a few NGC 6522 field stars to have $\Delta S \simeq 3$, implying [Fe/H] = −0.7. This is pretty metal-poor!

Another project is to study selected halo fields (Kinman *et al.*, 1965) with RR Lyrae at varying distances from the plane and the galactic centre. The ultimate aim would be to directly map the metal-abundance gradient in the halo. The NGP field is under investigation now; the field 'below the anti-centre' ($\alpha \sim 2^h$, $\delta = +40°$) reaches RR Lyraes nearly 20 kpc away (and 10 kpc down), and so far the most metal-poor stars have only [Fe/H] = −1.5 (surprisingly high). This may be contrary to the Eggen, Lynden-Bell and Sandage picture. Why are not there stars as metal-poor as in M92?

The giants in globular clusters are attractively bright stars, and have been studied by several observers. However, some caution is necessary − work by Zinn (1973) on the G-band in subgiants and asymptotic branch giants in M92 and Kraft's recent digital spectra of M92 stars strongly suggests changes in the molecular bands of CH, CN and NH as the star evolves up the subgiant branch, reaches the red giant tip and goes back on the asymptotic giant branch of the HR diagram. There is good observational evidence in M92 that C and N are anti-correlated, mixing starts early, and C^{12} is gradually processed into N^{14}. So metal-abundances by photometry, measuring CN, could be pretty hazardous; luckily there is no indication of any change in *heavy* elements in the M92 stars (Fe, Ca, Na, etc, seem quite low throughout).

Some red giants in the Draco system (the nearest dwarf E galaxy dynamically bound to the Milky Way) have been observed by Canterna (1975) and by Hartwick and McClure (1974). Both suggest [Fe/H] for Draco is lower than that found for M92; an average of their scattered abundances suggests [Fe/H] $\doteq$ −2.9, so these stars are really deficient in heavy elements − down a thousand times from a solar composition!

Draco is 68 kpc away now, and its tidal radius suggests that Draco couldn't have passed very close to the centre of our galaxy − it's always been way out in the halo, but is probably bound to the galaxy. If other distant systems like Sculptor or Pal 3 or Pal 4 also have extremely low abundances, there would be strong evidence for a halo abundance gradient extending from 10 to $\sim$100 kpc from the centre of our galaxy.

Canterna (1975) comments that any metal gradient established far out into the galactic halo is in agreement with the ELS model and a rapid collapse time; Larson's (1974) models for a spherical galaxy undergoing inhomogeneous collapse might be relevant too. In that case the radial dependence of Z (given mainly by *Draco*) would rule against Larson's models A-D, which predict [Fe/H] $\sim$ constant with distance for $r \geqslant 10^3$ pcs. However the Draco result would be consistent with models $E \rightarrow I$ with a collapse time $\sim 10^9$ yr, and a relatively *slow* and *low* star formation rate, which gives less metal production by stellar evolution in the early stages of the collapse when the stars of the outermost regions are formed.

Before we generalize too far from one beautiful, timely, pair of photometric observations, we should recall that the metallicity levels and gradients in globular clusters around other galaxies are probably different from our own. The Magellanic Cloud globulars are rather metal-poor. In the halo and disk of M 31 the known globular clusters are generally much more metal-rich than in our system (van den Bergh, 1969; Spinrad and Schweizer, 1972; Christensen, 1972), while the red colour of the M 87 globular clusters (Ables *et al.*,

1974) suggests they may be even systematically higher in heavy element abundances (cf. van den Bergh's 1975 review). I end this section with a query; much of our knowledge of the *halos* of other galaxies and our own comes from the study of globular clusters. Are these high density halo condensations typical of the surrounding sparse environs?

5. Abundances in Other Galaxies

Here we try to push to great distances and are often looking for rather subtle abundance effects. It is a challenge! Astronomers have begun to look at the abundances in other galaxies in a variety of complementary ways. On a small scale we can study individual luminous stars in the nearby Magellanic Clouds by conventional (but difficult) spectroscopy and comparison to (known?) galactic supergiants (cf. Fry and Aller, 1975; Przybylski, 1972; and Osmer, 1973).

Generally these spectroscopists found lower metals in both the LMC and the SMC, although the uncertainties are rather large. Van den Bergh (1975) and Gascoigne (1974) have suggested that classical Cepheids, whose mean period decreases from 6.1 days in the galaxy to 4.3 days in the LMC to 2.6 days in the SMC, indicate a lowering of the mean (internal) metal abundance in these *young* stars. Indeed, it appears as if stars of *any age*, apparently, have a low metal-abundance in the Magellanic Clouds. Cepheids' periods are an indirect, and easily-used, tool which should be also applied to abundance *gradients* in galaxies with a young-star population.

Most of the population/abundance work in other galaxies has been based upon integrated stellar spectra for the older-star dominated regions (nuclei of all types, bulges of Sa, Sb and SD galaxies, all over *E* galaxies) or in the interstellar medium (unborn new star population) – using H II region emission lines.

The analysis of the starlight shows line and band gradients in intermediate band photometry or low-resolution spectroscopy of the central parts of Sb-*E* galaxies. The CN bands have been fairly thoroughly studied by McClure (1969); Spinrad and Smith and Taylor (1972) and Spinrad and Stone (1975). That is because the λ 4200 CN band is wide and strong and relatively easy to measure in regions of low galaxies surface brightness.

Resually show a marked decrease in the CN [and when measured (cf. Welch and Forrester, 1972; Joly and Andrillat, 1973) strong neutral lines of Na, Fe and Mg] away from the nuclei of the galaxies. The scale length for 1/2 change from nucleus to a very low value, however, is variable – being as small as 100 pc in M 31 and 300 pc in N 3379 (EO). However, this could be deceptive, because of the limited nuclear angular resolution at 10–20 Mpc distance for giant *E* systems.

Spinrad *et al.* found variations in the CN gradients which correlated with the galaxies' radial drop in surface brightness. The *E* systems, which follow Hubble's law, generally show the steepest radial decline in CN, while the major axis of disk systems (Sa, SO galaxies) show little drop in metallicity with *r*. The SO system NGC 3115 has both populations; we therefore feel that the CN variation is associated with the old nuclear and halo populations, rather than the old disk. However that conclusion cannot be true for all E galaxies or most spirals, as recent Tololo CN measures by Spinrad and Stone seem to indicate a metal-rich halo to the Fornax EO galaxy NGC 1399 (halo CN $\approx$ nuclear value). Also as we note next, most spirals with gas and young stars show a sharp radial gradient in at least oxygen and nitrogen – from their H II region spectra.

Following the pioneering work of Peimbert (1968) on light-element abundances in the *nuclei* of M 81 and M 51, Searle (1971) made an extensive study of giant H II regions in several spiral galaxies. He confirmed the older Aller (1942) effect; the ratios of [N II]/Hα, Hβ/[O II] and [N II]/[O II] *decreased* from the centre outwards forming a neat, one-parameter family. The correlation holds from the inner H II regions (generally $\approx$ 1 kpc from the nucleus) to the outermost regions. Searle concluded that these spectroscopic changes were most likely due to the presence of both O/H and N/H abundance gradients with the O^{16}, N^{14}/H ratios increasing inwards. We should remark that these results are only slightly model-dependent. This trend, if extrapolated to the nuclear regions (also see Warner, 1973) would agree with Peimbert's (1968), 1975) approximate factor of 6–12 overabundance of N, in particular. Recently H. E. Smith (1975) has obtained new observations of the H II regions of M 101 and M 33. He found a strong radial gradient in the O/H ratio, which decreased a factor of $\sim$10 across these two galaxy disks. The Ne/C and S/O abundance ratios are constant. The nitrogen abundances indicate a similar but weak gradient in the ratio of N/O ($\sim$4 x over thehe disk).

TABLE V

Survey of selected Globular Cluster Abundances

Cluster	[Fe/H]$_{Canterna}$ indices	[Fe/H]$_{DDO}$	[Fe/H]$_{\Delta S}$
M92	−2.2 ± .4	−2.0	−2.2
M22	−2.0 ± .2		−1.7
M10	−1.7	−1.4	
M53	−1.6		−1.85
M3	−1.4	−1.0	−1.6
M71	−0.3 ± 0.2		−0.04

So for Sc galaxies all of this fits together pretty well – SMR nuclei-metal-rich inner regions – normal abundances $\approx$ 0.5 Holmberg radius, and metal-poor exteriors!

Complications exist here too – galaxies with big bulges or less gas (H I/ star ratio lower), seemed to have H II regions with systematically stronger [N II] lines – in accord with the Burbidge's (1962, 1965) previous result for nuclei.

Finally, I'd like to discuss the evidence for a metal-abundance dependence on galaxy type and/or mass.

It is pretty clear now, that for *E* galaxies – with only old stars – that abundance at the nucleus and away from it depends on total galaxy mass. We go from the little dwarf spheroidal Draco ($[Fe/H] \simeq -3$) to M 32 (nucleus $\approx$ solar) to the giant NGC 4472 ($[Fe/H] \approx +0.3$). These stellar determinations are upheld by the work of Jenner *et al.* (1973) who found $[O/H] \sim -1$ in the planetary nebulae of NGC 185, a dwarf E comparison to M 31.

Even better established are the gaseous composition differences between H II regions in small irregular galaxies (II Zw40, SMC) and our galaxy (near the Sun).

Dufour (1975), Aller *et al.* (1974) and Peimbert and Torres-Peimbert (1975) have observed H II regions in the Small Magellanic Cloud. I have averaged the well-determined ionic species (N, O, Ne) observed by these three sets of observers to form a mean comparison of the abundances of some light elements in the SMC, LMC and Orion-Carina of our Galaxy. Table VI lists these abundances.

This means the Small Cloud star-formation regions are down by 20 times in N, 6 times in O and 4 times in Neon! The overall metal abundance crudely-integrated over the

TABLE VI

Total Chemical Abundances in the H II regions[*]

Galaxy	N	O	Ne
⟨SMC⟩	6.42	8.04	7.31
⟨LMC⟩	7.10	8.58	7.94
(Ori + Carina) in our galaxy	7.64	8.80	7.99

[*] Tabulated is log $N(X)$ with $H = 12.00$.

periodic table, is about 1/5 solar! This seems to be well established.

It does seem attractive to try to understand overall metal abundances and gradients in terms of theoretical stellar and galactic evolutionary models. The work of Talbot and Arnett (1973, 1975) for example, does indeed predict gradients in primary (O^{16}) and secondary (N^{14}) nuclei over disks of galaxies. The radial distribution of H I, H II regons and of colours is reproduced — so that we could infer that small, low mass, H I-rich irregulars like the SMC (or even the Markarians or II Zw40) are like the outer extremes of big spirals — and thus should be metal-poor. The star-evolution yield of heavy elements just has been too small (up to now).

Finally, I would like to close with a remark on the He-problem. This major abundant element has hardly been mentioned in my talk, or indeed as an effect on spectral classification [except for a few peculiar B stars] (but see Nissen, this Symposium).

I personally have had the prejudice of 'everywhere-constant Helium'; i.e., $Y = 0.28 - 0.30$. But the new and accurate spectrophotometry by the Peimbert's for the low-metal H II regions in the LMC and the SMC, are very suggestive of somewhat lower He, too. Peimbert and Torres-Peimbert (1975) suggest $Y = 0.24$ in the SMC; they propose a correlation of He and metals in galaxies.

This in turn, implies a primordial, or rather, *pregalactic,* He abundance rather lower than supposed to originate in the 'Big-Bang'. The pregalactic $N(He)/N(H)$ is 0.074 ± 0.006, or $Y = 0.228$. So our galaxy and sun have benefited from star-produced He, too. This will, of course, have some considerable effect on our ideas of cosmology (an open universe if $H_o > 20$ km s^{-1} Mpc^{-1}) and the luminosity of young galaxies! It definitely requires independent confirmation.

As you can see, the subject of abundances in stellar populations is still an open and active field, and one that interacts with a variety of other astronomical and physical disciplines.

Acknowledgement

I wish to thank Drs. R. P. Kraft, B. J. Taylor, H. E. Smith, R. Canterna, and M. Raff for informative discussions.

References

Ables, H. D., Newell, B., and O'Neil, E. J.: 1974, *Publ. Astron. Soc. Pacific* **86**, 311.
Aller, L. H.: 1942, *Astrophys. J.* **95**, 52.
Aller, L. H., Czyzak, S. J., Keyes, C. D., and Boeshaar, G.: 1974, *Proc. Nat. Acad. Sci. USA.*, **71**, 4496.
Arp, H. C.: 1965, *Astrophys. J.* **141**, 43.
Barry, D. and Cromwell, R.: 1974, *Astrophys. J.* **187**, 107.
Becker, W. and Steinlin, U.: 1956, *Z. Astrophys.* **39**, 188.
Blanc-Vaziaga, M. J., Cayrel, G., and Cayrel, R.: 1973, *Astrophys. J.* **180**, 871.
Bond, H.: 1970, *Astrophys. J. Suppl.* **22**, 117.
Boyle, R. J. and McClure, R. D.: 1975, *Publ. Astron. Soc. Pacific* **87**, 17.
Burbidge, E. M. and Burbidge, G. R.: 1962, *Astrophys. J.* **135**, 694.
Burbidge, E. M. and Burbidge, G. R.: 1965, *Astrophys. J.* **142**, 634.
Butler, D. and Kraft, R. P.: 1975, in *Albany Photometric Conference* (in press).
Butler, D.: 1975, *Astron. J.* (in press).
Canterna, R.: 1975, in press.
Christensen, C.: 1972, unpubl. Caltech Ph.D. Thesis.
Dearborn, D. S., Eggleton, P., and Schramm, D. N.: 1975, preprint.
Dufour, R.: 1975, *Astrophys. J.* **195**, 315.
Eggen, O. J., Lynden-Bell, D., and Sandage, A. R.: 1962, *Astrophys. J.* **136**, 748.
Fry, M. A. and Aller, L. H.: 1975, *Astrophys. J. Suppl.* **29**, 55.
Gascoigne, S. C. B.: 1974, *Monthly Notices Roy. Astron. Soc.* **166**, 25P.
Grenon, M.: 1973, in G. Cayrel de Strobel and A. M. Delplace (eds.), 'Ages of Stars' *IAU Colloq.* No. 17, Obs. de Paris, Meudon.
Griffin, R.: 1975, *Monthly Notices Roy. Astron. Soc.* **171**, 181.
Gustafsson, B., Kjaergaard, P., and Anderson, S.: 1974, *Astron. Astrophys.* **34**, 99.
Harmer, D. L., Pagel, B., and Powell, A. L. T.: 1970, *Monthly Notices Roy. Astron. Soc.* **150**, 409.
Hartwick, F. D. A., Hesser, J., and Hill, G.: 1972, *Astrophys. J.* **174**, 573.
Hartwick, F. D. A. and McClure, R. D.: 1974, *Astrophys. J.* **193**, 321.
Hearnshaw, J.: 1972, *Mem. Roy. Astron. Soc.* **77**, 55.
Hearnshaw, J.: 1973, in G. Cayrel de Strobel and A. M. Delplace (eds.) 'Ages of Stars' *IAU Coll.* No. 17, Obs. de Paris, Meudon.

Hearnshaw, J.: 1974, *Astron. Astrophys.* **30**, 203.
Hearnshaw, J.: 1975, *Astron. Astrophys.* **38**, 271.
Helfer, H. L., Wallerstein, G., and Greenstein, J. L.: 1959, *Astrophys. J.* **129**, 700.
Janes, K.: 1972, Yale University Thesis.
Janes, K. A. and McClure, R. D.: 1971, *Astrophys. J.* **165**, 561.
Janes, K. A. and McClure, R. D.: 1973, in G. Cayrel de Strobel and A. M. Delplace (eds.), 'Ages of Stars', *IAU Colloq.* No. 17, Obs. de Paris, Meudon.
Jenner, D. C., Ford, H. C., and Epps, H. W.: 1973, *Bull. Am. Astron. Soc.* **5**, 13.
Johnson, H. L. and Knuckles, C. F.: 1955, *Astrophys. J.* **122**, 209.
Johnson, H. L., Mitchell, R., and Iriarte, B.: 1962, *Astrophys. J.* **136**, 75.
Joly, M. and Andrillat, Y.: 1973, *Astron. Astrophys.* **26**, 95.
Jones, D. H. P.: 1971, *Monthly Notices Roy. Astron. Soc.* **154**, 79.
Jones, D. H. P.: 1973, *Astrophys J. Suppl.* **25**, 487.
Kinman, T. D.: 1959, *Monthly Notices Roy. Astron. Soc.* **119**, 538.
Kinman, T. D., Wirtenan, C., and Janes, K.: 1965, *Astrophys. J. Suppl.* **11**, 223.
Larson, R. B.: 1974, *Monthly Notices Roy. Astron. Soc.* **166**, 585.
Lindblad, B.: 1922, *Astrophys. J.* **55**, 83.
Mayall, N. U.: 1946, *Astrophys. J.* **104**, 290.
McClure, R. D.: 1969, *Astrophys. J.* **74**, 50.
McClure, R. D.: 1973, In C. Fehrenbach and B. E. Westerlund (eds.), *IAU Symp.* **50**, 162.
McClure, R. D., Forrester, W. J. and Gibson, J.: 1974, *Astrophys. J.* **189**, 409.
Morgan, W. W.: 1956, *Publ. Astron. Soc. Pacific* **68**, 509.
Oinas, V.: 1974, *Astrophys. J. Suppl.* **27**, 391.
Osborn, W.: 1973, *Astrophys. J.* **186**, 725.
Osmer, P.: 1973, *Astrophys. J.,* **184**, L127.
Pagel, B.: 1972, in G. Cayrel de Strobel and A. M. Delplace (eds.) 'Ages of Stars', *IAU Colloq.* No. 17, Obs. de Paris, Meudon.
Pagel, B.: 1974, *Monthly Notices Roy. Astron. Soc.* **167**, 413.
Pagel, B. and Patchett, B. E.: 1975, *Monthly Notices Roy. Astron. Soc.* **172**, 13.
Peimbert, M.: 1968, *Astrophys. J.* **154**, 33.
Peimbert, M.: 1975, *Ann. Rev. Astron. Astrophys.* **13** (in press).
Peimbert, M. and Torres-Peimbert, S.: 1975, *Astrophys. J.* (in press).
Peterson, R.: 1975, *Astrophys. J.* (in press).
Preston, G. W.: 1959, *Astrophys. J.* **130**, 507.
Preston, G. W.: 1961, *Astrophys. J.* **134**, 651.
Przybylski, A.: 1972, *Monthly Notices Roy. Astron. Soc.* **159**, 155.
Raff, M.: 1975, private communication and U.C. Berkeley thesis.
Robinson, L. B. and Wampler, E. J.: 1972, *Publ. Astron. Soc. Pacific* **84**, 161.
Schwarzschild, M. and Schwarzschild, B.: 1950, *Astrophys. J.* **112**, 248.
Searle, L.: 1971, *Astrophys. J.* **168**,, 41.
Smith, H. E.: 1975, *Astrophys. J.* (in press).
Sneden, C.: 1973, *Astrophys. J.* **184**, 839.
Spinrad, H.: 1973, *Astrophys. J.* **183**, 923.
Spinrad, H. and Schweizer, F.: 1972, *Astrophys. J.* **171**, 403.
Spinrad, H. and Stone, R. P. S.: 1975, *Astrophys. J.* (in press).
Spinrad, H. and Taylor, B. J.: 1969, *Astrophys. J.* **157**, 1279.
Spinrad, H., Smith, H. E., and Taylor, D.: 1972, *Astrophys. J.* **175**, 649.
Strom, S. E., Strom, K., and Carbon, D.: 1971, *Astron. Astrophys.* **12**, 177.
Sturch, C.: 1966, *Astrophys. J.* **143**, 774.
Sturch, C.: and Helfer, L.: 1972, *Astron. J.* **77**, 726.
Talbot, R. J. and Arnett, W. D.: 1973, *Astrophys. J.* **186**, 69.
Talbot, R. J. and Arnett, W. D.: 1975, *Astrophys. J.* **197**, 551.
Taylor, B. J.: 1970. *Astrophys. J. Suppl.* **22**, 177.
van den Bergh, S.: 1969, *Astrophys. J. Suppl.* **19**, 145.

van den Bergh, S.: 1972, *Publ. Astron. Soc. Pacific* **84**, 371.
van den Bergh, S.: 1975, *Ann. Rev. Astron. Astrophys.* **13**, 217.
Wallerstein, G., Greenstein, J. L., Parker, R., Helfer, H. L., and Aller, L. H.: 1963, *Astrophys. J.* **137**, 280.
Warner, J. W.: 1973, *Astrophys. J.* **186**, 21.
Welch, G. and Forrester, W. T.: 1972, *Astron. J.* 77, 333.
Williams, P. M.: 1971a, *Monthly Notices Roy. Astron. Soc.* **153**, 171.
Williams, P. M.: 1971b, *Monthly Notices Roy. Astron. Soc.* **155**, 215.
Williams, P. M.: 1974, in *The 1st European Astronomical Meeting,* Athens: Springer-Verlag, Berlin, **2**, 174.
Williams, P. M.: 1974, *Monthly Notices Roy. Astron. Soc.* **167**, 359.
Zinn, R.: 1973, *Astrophys. J.* **182**, 183.

DISCUSSION

Griffin: In narrow-band photoelectric work it is generally appreciated that you measure *something* – that is, that you get numbers out of your machine but you are not always certain what the numbers mean. Although Arcturus was not included in the Spinrad-Taylor scanner studies, could you not measure it and compare your numbers with, say, the *Arcturus Atlas*? I am thinking in particular of the MG b-band region. Even in Arcturus, where Mg is known to be deficient, there is a surprisingly large number of strong MgH lines throughout that b-band region upon which you have based your Mg index. Incidentally, the strength of MgH may not necessarily be directly related to the Mg abundance, due to a variety of atmospheric effects upon the MgH dissociation equilibrium.

Spinrad: (1) It is too bright for the Lick equipment, unfortunately. (2) Maybe the MgH matters, but I doubt if it would vary out-of-phase with Mg I '*b*' in stars of similar temperature. Recall, the technique is differential.

Jaschek: Could you please quote average errors for the results you quoted? I am slightly surprised by the discussion of gradients of the order of $\Delta[\frac{Fe}{H}] = 0.023$ kpc^{-1} since the errors of individual determinations are of the order of some tenths.

Spinrad: The errors are *internally* small in the DDO δS(CN) measures. The gradient is that of the *mean*; the dispersion is, of course, substantial.

Steinlin: There is no definite limit between disc stars of varying metal content on one side and a halo population on the other side, but a continuous distribution change with concentration to the plane of the disc strongly correlated with metal content. This means that lines of equal space density of stars go from extremely flat ellipsoids for high metal content to more and more spherical distribution for low metal content. The difficulty in observing the necessary large number of faint stars today is that broad-band systems are not well suited for the differentiation one has to be able to make and some new kind of system has to be found for this work.

Spinrad: The boundaries suggested in my talk are completely artificial.

Bidelman: It appears to me that we now have both direct and indirect evidence of appreciable nucleosynthesis of nitrogen in giant stars: the C^{13}/C^{12} ratios and the nitrogen-gradients found in galaxies. Thus it seems almost certain that the CN strengths measured in red giants are partly the result of primordial abundances and partly due to effects of evolution. I therefore feel rather strongly that one should not attempt to infer Fe/H ratios from observed strengths of CN. This is especially so in the case of the CN-rich stars.

Williams: I think any interpretation of the ΔS of RR Lyrae stars in terms of [Fe/H] should take into account the systematic dependence of [Ca/Fe] with [Fe/H] observed both in narrow-band and many spectroscopic analyses.

Spinrad: Yes, Butler tried to do this.

Walborn: Some caution may be required in deriving abundances from low-resolution observations of giant and supergiant H II regions such as the Carina Nebula and 30 Doradus. High-resolution observations show complex line profiles which vary with position, as well as regions of greatly different excitations. These effects may average out, but that should be checked with more detailed observations. Alternatively, less spectacular H II regions should be preferred for abundance studies.

Spinrad: OK; however some observations of different regions of large H II regions (with differing line strengths) by Peimbert and by Smith, each yield similar light-element abundance ratios.

Cayrel: I do not think that the enhancement of Na, Mg and Ca which we found in some 'Spinrad's SMR' stars could be explained by a lowering of surface temperature in these stars caused by CN-molecular absorption, because the overabundances of Na, Mg and Ca persist even if one uses only lines of higher excitation potential, formed in much deeper atmospheric layers.

Morgan: In the case of the globular clusters, how many individual stars would have to be observed with high dispersion to simulate the average characteristics of metallicity of the cluster as a whole?

Spinrad: I think for the Fe, Cr group of relatively heavy metals, a few stars scattered over the HR diagram would be sufficient.

For CNO elements, I fear the worst – it may take much observing of stars all over the subgiant, giant and asymptotic giant branches. Kraft is now working on this problem.

Bell: From nucleosynthesis calculations, it seems to be easier to get stars with high N and low C^{13} rather than high N and high C^{13}. This may explain why μ Leo, which must have enriched N, has $C^{12}/C^{13} = 18$, whereas Arcturus doesn't seem to have enriched N and has a lower C^{12}/C^{13} ratio.

Furenlid: You questioned the validity of Peterson's conclusions regarding the SMR phenomenon because the three stars investigated by her are all strong line stars. But is not the point of her paper that a range in SMR characteristics does not correspond to a range in [Fe/H], independent of the level of [Fe/H] in these stars?

Spinrad: Partly true, but indeed the *photometric range* of [Fe/H] or δCN or some other line strength parameter in between the 3 giant stars is small – *none* are normal, all have photometric abundance excesses. A bad choice of standard stars, I fear. However, Peterson's basic conclusion still could be correct.

Gustafsson: I would like to moderate your statement concerning the difficulties to reconcile the results for 'SMR' stars obtained by Kjaergaard, Andersen and me using very narrow-band photometry and those from conventional spectroscopy. The reason for this is the fact that there *are* lines from the flat part of the curve of growth within our abundance index for the most metal-rich stars, just because the photometric system was not mainly designed for studying the SMRs but giants in general. Therefore, a considerable surface cooling or a somewhat abnormal microturbulence could bring our [Fe/H] values for the SMRs down to the order of [Fe/H] for the Hyades. One way to clarify the situation would be to use the very-narrow-band technique with groups of still weaker lines if suitable groups can be found and the lines can be identified.

Spinrad: I agree!

Keenan: The M-dwarfs are being re-classified by Mrs Boeshaar at Perkins, and our plates show that Kapteyn's star is a subdwarf with very *strong* lines, particularly Ca 4226.

Spinrad: That's very important, and makes Kapteyn's star roughly similar in spectrum and M_v and motion to Barnard's star. Thanks.

Payne-Gaposchkin: Will you say something about planetary nebulae as tracers?

Spinrad: They have an excellent potential in this role, especially in the galactic bulge and halo. Quantitative spectra are now needed.

Stephenson: I have a brief comment plus a totally unrelated and somewhat awful question. *The comment*: I believe it is correct to assert, as a purely historical point, that the first published claim that there exists massive evidence for a nitrogen deficiency in the Small Magellanic Cloud was by Sanduleak.

The question: As many here are aware, Fred Hoyle has recently published a fascinating hypothesis to explain the solar neutrino deficiency, involving the survival of stellar nuclei from a previous generation of the Universe. If this is correct, it is highly relevant to the subject of stellar populations vs chemical abundances. Did you, in the course of the vast thinking that you did for this very excellent review, consider what you would have said about Hoyle's theory had you decided to include it?

Spinrad: (1) OK, Sanduleak's work on planetaries in the SMC did suggest a very low N abundance.
(2) I don't know enough about Hoyle's idea – sorry!

McCarthy: A question and a comment.

Question: Do you know whether the sodium line observations reported in your paper have been corrected in each case for the effects of Na in the night sky?

Comment: It seems important for us in discussions of galactic structure at high latitudes not to confuse the local structure in the direction of the NGP and SGP with much more extensive (and involved) region which is the galactic halo itself. We should try not to confuse a small part with a larger whole.

Spinrad: Yes.

A POPULATION DISCRIMINANT IN M-DWARF SPECTRA

W. P. BIDELMAN

Case Western Reserve University, East Cleveland, U.S.A.

and

W. G. SMETHELLS

University of Wisconsin, Eau Claire, U.S.A.

Abstract. The CaH bands in the λ 6385 region of the spectra of M dwarfs have long been known; they become visible in the late K's and show a gradual increase in strength to later types. However, Greenstein and Eggen have noted that there is a substantial strengthening of all the hydride bands, including CaH, in high-velocity dwarfs (notably G95-59), and one of us (W.P.B.) has also seen this effect in suitable Warner and Swasey Observatory objective-prism plates of the rather low-velocity subdwarfs Ross 730 and 731. In a recent southern-hemisphere survey with the Curtiss Schmidt, Smethells has discovered nearly 200 late K and early M dwarfs, for all of which estimates of CaH-band strength were made. He found a considerable variation of the CaH-band strength in stars of the same (TiO-band) spectral type, and a slight tendency for the CaH/TiO band ratio to increase with increasing tangential velocity. A fuller account of this work will appear elsewhere.

DISCUSSION

McCarthy: My congratulations on the possible new metal abundance discriminant reported here. There are two other luminosity criteria available as discriminants, Na D lines (as discovered by W. Luyten in 1923) and MgH. Can you tell us if any correlations have been made with these?

Bidelman: CaH and MgH behave very similarly. As far as the D lines are concerned, we have used their strength as the primary discriminator between giants and dwarfs, but in view of their confusion with a strong TiO band I would think it difficult to use these for any other purpose.

B. Hauck and P. C. Keenan (eds.), Abundance Effects in Classification, 205. All Rights Reserved.
Copyright © 1976 by the IAU.

CHEMICAL EVOLUTION OF THE GALACTIC DISK AND THE RADIAL METALLICITY GRADIENT

M. MAYOR

Observatoire de Genève, Switzerland

Abstract. An analysis of the kinematical and photometric properties of about 600dF stars and 600 gG-gK stars permits the estimation of the radial chemical gradient in the Galaxy. The mean value in the solar neighbourhood obtained for all of these stars is:

$$\frac{\partial[\text{Fe/H}]}{\partial\varpi} = -0.05 \pm 0.01 \text{ kpc}^{-1}$$

The values of [Fe/H] used for this estimation are deduced for the dF stars using $uvby\,\beta$ photometric measurements and for the gG-gK stars from a list published by Hansen and Kjaergaard. An estimate of the chemical gradient using UBV photometry of dG stars in the solar neighbourhood gives a similar value. For all the samples studied (dF, dG or giants) the order of magnitude for the gradient is the same. However, for the youngest stars in these samples the metallicity gradient could be larger:

$$\frac{\partial[\text{Fe/H}]}{\partial\varpi} = -0.10 \pm 0.02 \text{ kpc}^{-1}$$

Such a value may be affected by dynamical perturbations of the galactic disk.

The values published by Hansen and Kjaergaard for the sodium concentration in giant star atmospheres also indicate a radial galactic gradient of the same order.

If only the dF stars which are sufficiently evolved to allow an age estimate are considered, then a very distinct correlation is found between age and metallicity:

$$\frac{\partial \bar{Z}/\bar{Z}_\odot}{\partial t} = 0.6 \pm 0.3/10^{10} \text{ yr}$$

An important fraction of the heavy elements actually present in the solar neighbourhood seems to have synthetized during the life of the galactic disk.

The two derivatives $\dfrac{\partial \bar{Z}}{\partial t}$ and $\dfrac{\partial Z}{\partial \varpi}$ are not independent, but are connected by the chemical evolution of the galactic disk. Some elementary deductions show the coherency of these two estimates.

The intrinsic dispersion of metallicities, at a given age and birthplace, is somewhat lower than the admitted values.

$$\sigma_{[\text{Fe/H}]}\big|_{\varpi,\,t} = 0.10 \left(\begin{array}{c} +0.05 \\ -0.10 \end{array}\right)$$

B. Hauck and P. C. Keenan (eds.), Abundance Effects in Classification, 207–208. All Rights Reserved.
Copyright © 1976 by the IAU.

It has not been possible to find any significant variation with age of this quantity from the present observational material. The simultaneous variation of σ_w^2 and [Fe/H] as function of age is evidence for a z stratification in the mean abundance of the heavy elements. The ratio between the mean metallicity in the plane and at $z = 500$ pc is estimated to be about a factor of two.

Finally it is shown that the interpretation of the kinematical diagrams for different groups of given metallicity is ambiguous. A relation as e vs [Fe/H] depends not only on the chemical and kinematical history of the Galaxy but is also strongly dependent on the observational errors of [Fe/H] and on criteria used to define the sample.

A paper containing the above results has been submitted for publication in *Astronomy and Astrophysics.*

DISCUSSION

Gerbaldi: Did you have some Fm stars in your statistic? In view that those stars are at the cool end of the Am phenomena and that the metallicity is produced by diffusion, i.e. in a quite short time regarding the evolution time, the presence of these stars could introduce some disturbance in your results.

Mayor: All stars with noted peculiarities in the B5 catalogue have been excluded but evidently it is possible that a small number of unrecognized Fm stars be present.

Bell: Do you plan to make use of Maeder's evolutionary tracks, which give a better fit to cluster main sequences?

Mayor: At the time of the beginning of this investigation the evolutionary tracks of Maeder were not available but I intend to use them.

ABUNDANCES IN THE GALACTIC
CLUSTERS OF HYADES AND M 67

R. FOY

Observatoire de Meudon, France

Abstract. Two Hyades dwarfs and two giants in the old galactic cluster M 67 have been carefully analyzed from 7 Å mm^{-1} taken at OHP with an electronic camera. Measurements of weak lines lead to ə the conclusion that both these clusters have a solar chemical composition, although they have very different ages.

To know the global metal abundance of Hyades stars is essential, at least for two reasons: firstly to derive an accurate age for this cluster from theoretical evolutionary tracks; secondly to have a reliable calibration of chemical composition for photometric or low resolution spectroscopic works. So abundance determinations of the Hyades are very numerous. I would want to add another one which will be convincing, I hope, that the metal content of the Hyades is not significantly different from the solar one.

I have studied the two solar type dwarfs of the Hyades cluster HD 28068 (=63) and HD 28344 (=73), from spectra taken by R. Cayrel at the Observatoire de Haute-Provence with an *échelle* spectrograph and a Lallemand electronic camera. The dispersion of the spectra is 6 Å mm^{-1}. Spectra of Vesta have been taken in the same way to check equivalent width measurements.

These two dwarfs have been analyzed using Peytremann's model atmospheres. From curves of growth well defined in their weak line part, it comes for the two stars the iron abundance $\left[\frac{Fe}{H}\right]_{\odot}^{Hyades} \simeq +0.05$, relative to the Sun.

This abundance is strongly confirmed by the diagram of Figure 1. HD 28068 has the same temperature as the Sun, from colours or from ionization equilibrium. Therefore, differences in equivalent widths of weak lines between HD 28068 and the Sun have to be interpreted in terms of difference in abundance. As shown in Figure 1, the systematic difference in equivalent widths between HD 28068 and Vesta is very small. This proves, by a way independent from atmosphere models, that the Hyades have a solar metal content: it is not possible to assume values of the metal content such as +0.20 or +0.30 for HD 28068. Note that adopting a slightly different effective temperature does not change this conclusion.

I have also analyzed eight giants belonging to the cluster or to the moving group of the Hyades. The mean abundance I have obtained for these giants: $< \left[\frac{Fe}{H}\right]_{\odot}^{Hyades} > = +0.06$ agrees very well with the abundance determined for the two Hyades dwarfs. This good agreement gives confidence in the reliability of abundances determined from the used model atmospheres of giants.

I have therefore applied them to study two giants in the old galactic cluster M 67. I have

B. Hauck and P. C. Keenan (eds.), Abundance Effects in Classification, 209–212. All Rights Reserved.

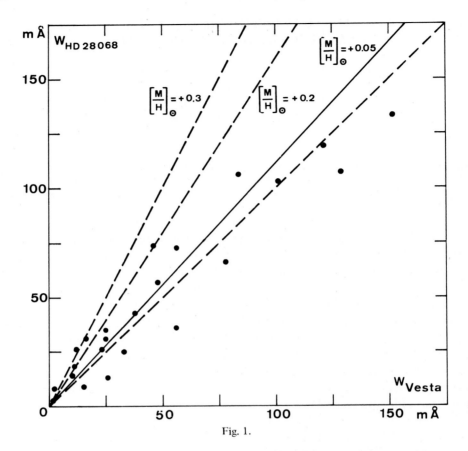

Fig. 1.

chosen them not too bright in order to avoid criticism about non-LTE effects. These stars are F 170 and F 244. I have taken five spectra at the Observatoire de Haute-Provence, with the *échelle* spectrograph and the Lallemand electronic camera at the Coudé focus of the 1.52 m telescope. The dispersion was 6 to 8 Å mm^{-1}.

From the well-defined curve of growth of F 170, I have found $\left[\frac{Fe}{H}\right]_{\odot}^{F170} \simeq -0.18$. Other elements have a similar slight deficiency.

The effective temperature of F 244 deduced from $R-I$ implies inconsistent results. Note that this star is misplaced in the colour-colour diagram of the DDO photometry. Therefore, I have determined its atmospheric parameters independently from the photometric data, drawing a polygonal diagram as described in Blanc-Vaziaga *et al.* (1973). A satisfying model could be $\theta_{eff} = 1.20$ and $\log g = 1.70$; the resulting abundance is $\left[\frac{Fe}{H}\right]_{\odot}^{F244} \simeq -0.02$.

So the mean abundance I find for these stars in M 67 is −0.10: let us say that M 67 has a solar metal content: from my analysis, it cannot be super metal-rich; the large absorption measured by Spinrad and Taylor (1969) in M 67 stars has to be interpreted in other terms than in terms of global overabundance of M 67 stars relative to the sun.

I would want to conclude by emphasizing implications of these analyses about the chemical evolution of our Galaxy. First, the rather young cluster of Hyades has a solar metal content. Second, the old galactic cluster M 67 has also a solar metal content. This would support the idea that there is no significant metal enrichment in the galactic disk since the birth of M 67, about 7.10^9 ago. This conflicts with the recently proposed models of chemical evolution of the galactic disk, such as that by Pagel or Tinsley.

References

Blanc-Vaziaga, M.-J., Cayrel, G., and Cayrel, R.: 1973, *Astrophys. J.* **180**, 871.
Spinrad, H. and Taylor, D. J.: 1969, *Astrophys. J.* **157**, 1279.

DISCUSSION

Williams: First, a comment: the mean Hyades giant star abundance you give includes abundances of several stars assigned to the Hyades *kinematic group* — whose membership is controversial. The mean abundance of the three Hyades *cluster* giants you give is [Fe/H] = 0.17, close to the canonical value.

Secondly, what damping do you use for the Mg *b* lines?

Foy: (1) I do not think that [Fe/H] = +0.17 is significantly different from [Fe/H] = +0.06.

(2) I use the damping constant resulting from the fit of the theoretical profile for the Sun to the observed one. This is therefore independent from the theory of damping.

Cayrel: Even if the Hyades giants have: [Fe/H] ~ +0.2 the important fact is that the Hyades dwarfs have: [Fe/H] ~ 0.

Osborn: You have given in your slides values of surface gravity and effective temperature for these stars. Have you checked these values by computing masses for these stars using the distance modulus? What are the masses you derive?

Foy: Yes, I have. Masses would lie between 0.2 and 0.8 or 1.0 $M_\odot$, depending upon the adopted distance modulus.

Martinet: In your statement on the non-enrichment of the galactic disc, you implicitly assume that NGC 188 and M 67 are typical, representative of objects of their age. On the contrary, one could consider these clusters as peculiar (long-lived amongst the open clusters) objects. Your statement may be correct but ought to be *proved* with the help of substantial material including objects well spread in age and in possible regions of formation. Further, NGC 2420 could be a counter-example.

Foy: I do not know if the dynamical peculiarity of M 67 can be related to the chemical composition of the interstellar matter which has generated it.

In our group in Meudon, we have analysed a lot of stars which are old and have a normal metal content.

It will be interesting to perform detailed analyses of stars in NGC 2420.

Gustafsson: You concluded from your results for the solar-type dwarfs that scaled Peytremann models may be used with confidence for the giants. I am not sure you are quite correct in that conclusion since Peytremann did not include molecular lines which are of some importance for the temperature structures of the giants (as is shown by our models) but not for the solar-type dwarfs.

Foy: We have investigated with G. Cayrel and T. Tsuji the region of the $\log g/T_{eff}$ plane where molecular blanketing affects the $T(\tau)$ law in the line formation optical depths (Foy, 1974, thesis).

Models used in the present study fall in the 'good' region of this plane (except, maybe, in the case of α Tau): effects of molecular blanketing do not significantly affect the derived abundances.

Obviously, if both atomic and molecular blanketed models, as your ones, had been available when I started this work, I would have used them.

Nissen: Your determination of the iron abundance of the two Hyades G dwarfs seems convincing, because you measured the equivalent widths of very weak lines, i.e. $W < 30$ mÅ. Some of the discrepancy with the higher abundance obtained from photoelectric photometry may be explained if the Hyades have a peculiar high 'microturbulence'. Do you find any indication of this in your *own* electronographic data?

Foy: I think that the discrepancy between photometric and spectroscopic determination of Hyades abundance could come from the influence of microturbulence on photometric bands.

In the present work, I have mainly studied weak lines, so, now, I have no indication of that from my data. But we are investigating this point with R. and G. Cayrel.

Bell: How do you explain the ultra-violet excess of the Sun relative to the Hyades main sequence, if the Sun has the same abundance as the Hyades?

Earlier suggestions by Conti and Deutsch that high δm_1 values could be produced by high Doppler Broadening Velocities have never been confirmed spectroscopically.

Foy: (1) Colours of the Sun are not known with enough accuracy to be really sure of this excess. It would be related to a higher microturbulent velocity in Hyades dwarfs.

(2) Accurate measurements of Doppler Broadening Velocities is very delicate in dwarfs, because of the large differences in damping constants – this is very well shown on my iron curve of growth of the Sun (Foy, 1972), of which the 'flat' part is strongly split.

Mayor: The difference that you obtain between Hyades and M 67 metallicities is 0.15 ± 0.30 (?) so from this, one can estimate a difference of about $\Delta[\text{Fe/H}] = 0.3 \pm 0.6$ in 10^{10} yr. It seems to me that this fact is not a proof against a galactic enrichment.

Foy: I do not say that my results prove that there is no metal enrichment in the galactic disc, but that they support this idea.

CURVE-OF-GROWTH ANALYSIS OF A RED GIANT IN M 67

R. GRIFFIN

Cambridge University, U.K.

Abstract. A coudé spectrogram of a red giant (IV-202) in M 67 has been analyzed with respect to Arcturus by the differential curve-of-growth method. The overall metal abundance is found to be approximately twice that in Arcturus, i.e. half the solar value. The error limits exclude the interpretation that IV-202 is 'super-metal-rich'. According to other observational evidence, IV-202 is quite typical of the red giants in M 67.

Reference

Griffin, R.: 1975, *Monthly Notices Roy. Astron. Soc.* **171**, 181.

DISCUSSION

Bidelman: Could you give us a more conventional or more respectable designation for this object?

Griffin: It is BD + 12° 1919.

Williams: Could you tell us a bit more about your determination of the temperature of this star – particularly whether you have any feel for the line temperature (e.g. $\Delta\theta_{ex}$) and how it compares with α Tau.

Griffin: First I derived a rough value of $\Delta\theta_{ex}$ from partial curves of growth. (I had to assume that differential excitation and effective temperatures would be identical). Then, accepting the value $\Delta\theta = +0.11$ for (α Tau – α Boo) given recently by van Paradijs and Meurs, I turned to Eggen's photometry and interpolated to find the corresponding differential temperature for IV-202.

Osborn: What is the approximate bolometric magnitude of this star?

Griffin: The value given by Eggen is −1.75.

Spinrad: Indeed, M 1465 = IV-202 in M67, from ST scans still has the visual M 67 anomaly-strong resonance lines of Ca, Mg, Na I. These lines are stronger than the normal field giant of a similar red colour.

Griffin: I believe the enhancement of such lines might be expected in a star that is both as cool and as luminous as IV-202; its estimated spectral type is K3 or K4, luminosity class II-III.

SOME COMMENTS ON THE AGE–ABUNDANCE RELATION

S. GRENIER

Observatoire de Paris, Meudon, France

L. DA SILVA

Observatorio Nacional, Rio de Janeiro, Brazil

and

A. HECK

Institut d'Astrophysique de l'Université de Liège, Belgium

Abstract. From the results obtained with the tri-dimensional quantative classification method of Spite (1966), the relation age-abundance is analyzed for the Galaxy. The dispersion on the abundance for stars at a given age appears to prevail over a possible relation between the age and the abundance for these stars. From a detailed analysis of eight F2-G2 stars, it appears that the index Δm_1 is sensitive to microturbulence velocity.

The method of quantitative tri-dimensional classification for stars from F5 to K0 elaborated by Spite (1966) was recalibrated (da Silva and Grenier, 1976) for two reasons. First, L. da Silva and F. Spite have performed new measures and, secondly, a lot of stars were observed after 1966 with high resolution and are now available in the literature, giving new stars usable as valid standards.

The main advantages of the method are the following. The derived results are independent of the interstellar reddening and of the microturbulence. Moreover the spectra used allow also the derivation of radial velocity (useful for the research of the relations between the kinematical and atmospheric parameters). The calibrated quantities are given in Table I.

TABLE I

Parameter	M_V	θ_{eff}	[Fe/H]	$(B-V)_0$
Standard deviation	0.37	0.015	0.17	0.014

The results obtained were used in order to search for a relation between metallicity and age (see Figure 1) only for the 197 stars, the evolutionary degree of which allowed a sufficiently reliable estimate of the age from the evolutionary tracks given by Iben (1967). The influence of metallic abundance determination of the age was not taken into account for no usable grid of isochrones was found.

18 stars were considered separately. They deviate significantly from the kinematical law of the total sample as was shown (see Grenier *et al.*, 1976) by the independent

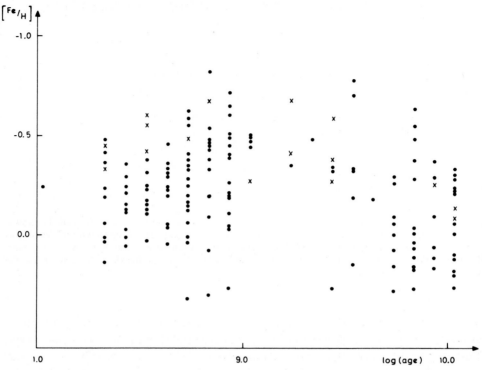

Fig. 1. Metallic abundance vs log (age) for the total sample. Crosses represent stars which deviate significantly from the kinematical law of the total sample.

application to this sample of a method of statistical parallaxes (see, e.g. Jung, 1970; Heck 1975a, b). These stars (shown by crosses) can be considered as not belonging to the galactic disk.

Figure 2 shows only the 75 stars observed by S. Grenier in the direction of the galactic rotation. They constitute an unbiased sample. In these two figures, no clear relation appears between metallicity and age.

It seems between 10^8 and 10^{10} yr, the dispersion of the abundances, for a given age, prevails over any relation with the age. This conflicts with the results found by Mayor (1975) who finds metal enrichment in the galactic disk, through the Δm_1 index of the $uvby\beta$ photometry of F stars. To explain this conflict two of us (LDS and SG) and R. Foy have analyzed the influence of the microturbulence on the [Fe/H] determination from the Δm_1 index. Two of us (LDS and SG) and R. Foy have analyzed the influence of the microturbulence in the [Fe/H] determination from the Δm_1 index of the $uvby\beta$ photometry.

Figure 3 shows an unexpected relation between Δm_1 and metallic abundance for dwarfs and subgiants studied with high resolution (~ 3 Å mm^{-1}) in the same way either by M. Spite (1967a, b, 1968) or by L. da Silva (da Silva, 1975; Scheid and da Silva,

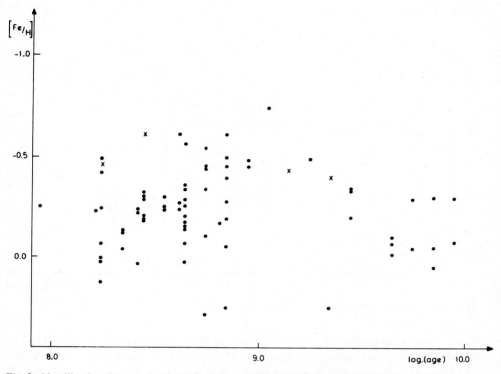

Fig. 2. Metallic abundance vs log (age) for 75 stars of an unbiased sample. Crosses have same meaning as in Figure 1.

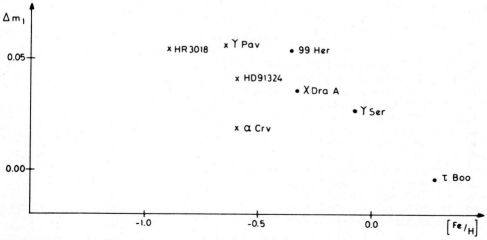

Fig. 3. Δm_1 vs metallic abundance for stars studied with high resolution by M. Spite (dots) and L. da Silva (crosses).

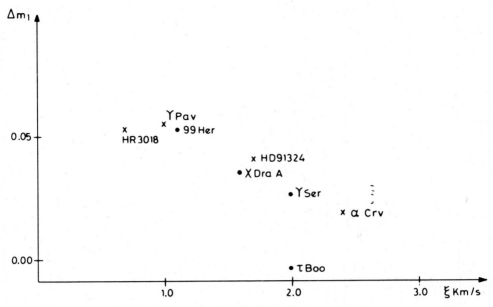

Fig. 4. Δm_1 vs microturbulence for same stars as in Figure 3.

1975). However, in Figure 4 it appears that Δm_1 is also a function of ξ. Three stars (γ Pav, HD 91324 and α Crv) analyzed in detail by da Silva have the same abundance, but have different Δm_1 in relation to their microturbulence velocities. Two other stars (γ Ser and τ Boo), analyzed by M. Spite, have the same ξ, but different Δm_1 because their [Fe/H] are different.

Thus it seems that Δm_1 is sensitive both to [Fe/H] and to ξ.

References

da Silva, L.: 1975, *Astron. Astrophys.* **41**, 287.
da Silva, L. and Grenier, S.: 1976 (in preparation).
Grenier, S., Heck, A., and Jung, J.: 1976 (in preparation).
Heck, A.: 1975a, thesis, Univ. Liège.
Heck, A.: 1975b, *Astron. Astrophys.* **43**, 111.
Iben, I.: 1967, *Ann. Rev. Astron. Astrophys.* **5**, 571.
Jung, J.: 1970, *Astron. Astrophys.* **4**, 53.
Mayor, M.: 1976, this Symposium, p. 207.
Scheid, P. S. and da Silva, L.: 1975 (in preparation).
Spite, F.: 1966, *Ann. Astrophys.* **29**, 601.
Spite, M.: 1967a, *Ann. Astrophys.* **30**, 211.
Spite, M.: 1967b, *Ann. Astrophys.* **30**, 685.
Spite, M.: 1968, *Ann. Astrophys.* **31**, 269.

REMARKS ON SOME ASPECTS OF SPECTRAL CLASSIFICATION

W. W. MORGAN

Yerkes Observatory, University of Chicago, U.S.A.

Abstract. The present state of MK spectral classification is described briefly.

1. Introduction

It is now over 30 years since the publication of the MKK *Atlas of Stellar Spectra*; and the time has come for a general re-examination of the mission of spectral classification. The principles outlined in the Introduction to the Yerkes Atlas of 1943 still seem valid; but the classification problem has become more complex in the intervening years through new astrophysical discoveries – and through the rapid growth of spectral classification by narrow-band photoelectric photometry (see, for example, B. Strömgren 1966, *Ann. Rev. Astron. Astrophys.* **4**, 433).

Such a re-examination is now in progress in two separate works: (1) the preparation of a revised atlas of spectral classification for the stars of later type, by Keenan; and (2) a new atlas of stellar spectra for stars earlier than the Sun, by Abt and Morgan. In the case of the latter work a major increase in the precision with which the MK system is defined is being achieved; this increase in systemic accuracy has already led to the discovery of several new categories of 'peculiar' stellar spectra.

An interesting example of a 'peculiar' category has been noted recently by Abt and Morgan for a group of metallic-line stars having rather strong K-line intensities. Members of this group, for which the Am star HD 103877 can be considered a prototype, have metallic-line types around F3-F5, and luminosity classes IV or V, in the neighbourhood of λ 4300. On the other hand, in the spectral range λλ 3850–4100 the luminosity classes range from Ib to III. In the case of the prototype, the spectrum in the latter region resembles closely that of α Persei (F5 Ib) on spectrograms of scale near 125 Å mm^{-1}, except for a weaker K line. Thus, we can say that the spectrum in the violet region noted does not correspond to that of a main-sequence star of any spectral type (Abt and Morgan: manuscript in preparation).

2. The Migration of Types over the Spectral Type – Luminosity Class Diagram

Some years ago, it was the custom to define spectral class O by stars whose spectra contain lines of He II. As the quality of spectrograms improved, more and more B0 stars were found to contain He II lines – and thus were moved into class O. Occurrences such as this emphasize the need for defining spectral types and luminosity classes in terms of standard stars. This is one of the principal features of the *MKK Atlas* of 1943.

B. Hauck and P. C. Keenan (eds.), Abundance Effects in Classification, 219–220. All Rights Reserved.

The need for some such procedure is not always realized at the present time; 'revised' MK types are still being published by some investigators in which the fundamental standards which define the MK system are given new values. Such procedures result in a floating classfication system, where absolute determinations of types result and the differential advantage of MK classification is lost. But this does not at all imply that the MKK or MK standards are sacrosanct; some of them have been changed in the past; and more are being changed in the present, in connection with the preparation of the new atlases. *But it is important to specify a new set of standard stars when the MK standards are altered.* If this is not done, the noise level is raised over the entire two-dimensional diagram.

3. The Mission of Spectral Classification
in the Near Future

Among other ongoing projects which are concerned with fundamental spectral classification, we note one of prime importance: the monumental *Michigan Catalogue of Two-Dimensional Spectral Types for the Henry Draper Stars*, where the principal classification is being carried out by Dr Nancy Houk. The first volume, which includes types from the South Pole to −53° declination, will emerge from press in the very near future. This project sets a new epoch in spectral classification and establishes its principal classifier as in the strain of such figures of the past as Miss Annie Cannon — and the greatest of all spectral classifiers, Miss Antonia C. Maury. The completion of the Michigan Catalogue is the most pressing of all the classification projects. It will furnish a wealth of new material for investigations of Galactic structure — and for astrophysical research. The greatest problem remaining for it is how the plates for the Northern fields are to be obtained.

The most important task for the revised MK atlases of Keenan and of Abt and Morgan is to improve the systemic precision to the greatest degree practicable. This will result in more accurate spectroscopic luminosities and parallaxes on one hand, and the identification and delineation of new categories of objects of astrophysical interest. Each atlas should be accompanied by an extensive catalogue of stars classified according to the new precepts — and, in effect, defining the revised system illustrated.

APPENDIX

A CATALOGUE OF [Fe/H] DETERMINATIONS

A CATALOGUE OF [Fe/H] DETERMINATIONS

M. MOREL

Institut d'Astronomie de l'Université de Lausanne
and Observatoire de Genève, Switzerland

and

C. BENTOLILA and G. CAYREL

Observatoire de Meudon, Paris, France

and

B. HAUCK

Institut d'Astronomie de l'Université de Lausanne
and Observatoire de Genève, Switzerland

Abstract. A catalogue resulting from the compilation of published values of iron/hydrogen abundances is given for 515 stars.

1. Introduction

The idea of presenting to Symposium No. 72 a catalogue of iron/hydrogen abundances coming exclusively from high dispersion analyses ($\leqslant 20$ Å mm^{-1}) came to us in remembering the high interest expressed in two previous lists of iron/hydrogen abundances, one published by R. and G. Cayrel (1966) and the other published by G. Cayrel *et al.* (1970). This time the authors have the pretension of calling the third list a 'Catalogue of [Fe/H] Determinations'.

2. The Catalogue

The catalogue consists of three tables.

In Table I, a compilation of absolute abundance determinations of iron in the solar photosphere made by Blackwell (1974) is given. These abundances are based on the scale $\log N_H = 12$.

In Table II, iron/hydrogen abundances of 515 stars are given. These abundances are given in the form of logarithmic differences between the iron abundance in the atmosphere of the analyzed star and the iron abundance in a standard star and written in the form:

$$\left[\frac{Fe}{H}\right]^{star}_{stand} = \log \frac{Fe}{H}(star) - \log \frac{Fe}{H}(stand)$$

B. Hauck and P. C. Keenan (eds.), Abundance Effects in Classification, 223–259. *All Rights Reserved.*
Copyright © 1976 by the IAU.

They come exclusively from detailed analyses based on high dispersion spectra. The chosen limit of the dispersion is about 20 Å mm^{-1}.

In Table III, indications on the mean dispersion and the wavelength interval used in each detailed analysis are given.

In addition to the abundance parameter, Table II contains other useful spectroscopic and photometric parameters, as the effective temperature gravity and microturbulence parameters with which $[\frac{Fe}{H}]^{star}_{stand}$ has been determined. Also a column indicating the standard star used in the $[\frac{Fe}{H}]^{star}_{stand}$ determination and a column for the bibliographic references have been added. The MK spectral type of the stars contained in Table II has been chiefly taken from the new Catalogue of Jaschek and Jaschek (1975). The photometric data have been compiled by the Stellar Data Center of Strasbourg.

3. Description of Table II

This table contains 515 stars and 876 measurements. It contains in brief the following columns:

Column 1:   HD number. Stars without an HD number are grouped at the end of the Catalogue

Column 2:   star name

Column 3:   spectral type taken chiefly from the new catalogue of Jaschek and Jaschek (in preparation). Spectral types given in brackets are taken from the 'Kennedy Catalogue' or other sources.

Column 4:   visual apparent magnitude VM taken briefly from the 'Bright Star Catalogue' or from the Catalogue of Jaschek et al. (1972) (magnitude VM followed by an asterisk).

Column 5:   absolute magnitude MV taken from the 'Gliese Catalogue'

Columns 6 and 7:   photometric data $B-V$, $U-B$ taken from Johnson and Mitchell (1966). The values from Jaschek et al. or other sources are followed by a double asterisk.

Column 8: $R-I$ (without an asterisk) taken from Johnson and Mitchell (1966) or, if followed by an asterisk, $I-K$ taken from Neugebauer and Leighton (1969)

Columns 9 and 10:   temperature parameter $\theta_{eff} = 5040/T$ with an indice (column 10):
(1) effective temperature
(2) ionization temperature
(3) excitation temperature
(4) effective temperature calculated by taking $\theta_{eff} = 0.87$ and assuming:
$$\Delta\theta_{exc} = \Delta\theta_{ion} = \Delta\theta_{eff}$$

Column 11:   gravity: $\log g$

Column 12:   electron pressure: $\log Pe$
Column 13: $[Pe]\,^{\text{star}}_{\text{stand}}$
Column 14: $[V]\,^{\text{star}}_{\text{stand}}$
Column 15:   microturbulent velocity
Column 16:   the standard star employed in this analysis
Column 17:   the iron/hydrogen parameter $[\frac{Fe}{H}]^{\text{star}}_{\text{stand}}$
Column 18:   reference

4. Conclusion

This catalogue may have omitted some available data of iron/hydrogen determinations, up to the epoch in which it was compiled (June 1975). The authors would be happy to receive such data, as well as new data of Fe/H determinations in order to prepare revised and updated editions of the Catalogue.

The Catalogue is dedicated to Prof. Chalonge, who has contributed in so many and highly interesting ways to spectral classification.

References

Blackwell, D. E.: 1974, *Quart. J. Roy. Astron. Soc.* **15**, 224.
Cayrel, R. and Cayrel de Strobel, G.: 1966, *Ann. Rev.* 4, 1.
Cayrel de Strobel, G., Chauve-Godard, J., Hernandez, G., and Vaziaga, M. J.: 1970, *Astron. Astrophys.* 7, 408.
Jaschek, C. and Jaschek, M. (in preparation).
Jaschek, C., Hernandez, E., Sierra, A., and Gerhardt, A.: 1972, *Obs. Astrophys. Nac. La Plata*, vol. 38.
Johnson, H. L. and Mitchell, R. I.: 1966, *LPL* **4**, 99.
Neugebauer, G. and Leighton, R. B.: 1969, The Two-Micron Sky Survey, *NASA SP-3047*.

TABLE I

Absolute abundance determinations of Fe with solar photosphere based on the scale $\log N_H : 12$

$\log N$ [Fe]		Publication
6.59	Fe I + Fe II	1964 (1)
7.63	Fe II	1969 (2)
7.62		1951 (3)
6.85	Fe I	1967 (4)
7.28	Fe I + Fe II	1972 (5)
7.60	Fe I	1969 (6)
6.57	Fe I + Fe II	1960 (7)
6.64	Fe I	1964 (8)
7.50	Fe II	1969 (9)
7.64	Fe I	1973 (10)
6.76	Fe I	1963 (11)
7.2	Fe I	1973 (12)
7.2	Fe I	1973 (13)
6.70	Fe I	1964 (14)
7.50	Fe II	1970 (15)
(5.8)		1925 (16)
7.55	Fe I + Fe II	1970 (17)
6.85	Fe I	1969 (18)
7.2	Fe I	1970 (19)
7.4	Fe I	1973 (20)
7.2	Fe I + Fe II	1929 (21)
6.78	Fe I	1964 (22)
7.15		1948 (23)
6.81	Fe I	1964 (24)
6.55	Fe I + Fe II	1968 (25)
6.45		1955 (26)
6.80	Fe I	1969 (27)

REFERENCES TO TABLE I

(1) Aller, L. H., O'Mara, B. J., and Little S.: 1964, *Proc. Nat. Acad. Sci.*, Washington **51**, 1238.
(2) Baschek, T., Garz, R., Holweger, H., and Richter, J.: 1969, *Astron. Astrophys.* **4**, 229.
(3) Claas, W. J.: 1951, *Rech. Astron. Obs, Utrecht* **12**, part I.
(4) Cowley, C. R. and Warner, B.: 1967, *Observatory* **87**, 117.
(5) Foy, R.: 1972, *Astron. Astrophys.* **18**, 26.
(6) Garz, T., Holweger, H., Kock, M., and Richter, J.: 1969, *Astron. Astrophys.* **2**, 446.
(7) Goldberg, L., Muller, E. H., and Aller, L. H.: 1960. *Astrophys. J. Suppl.* **5** no. 45.
(8) Goldberg, L., Kopp, R. A., and Dupree, A. K.: 1964, *Astrophys. J.* **140**, 707.
(9) Grevesse, N. and Swings, J. P.: 1969, *Astrophys. J.* **2**, 28.
(10) Huber, M. C. E. and Tubbs, E. F.: 1973, *Astrophys. J.* **186**, 1053.
(11) Leftus, V.: 1963, *Bull. Astron. Inst. Czech.* **14**, 155.
(12) Lites, B. W. and Brault, J. W.: 1973, *Solar Phys.* **30**, 283.
(13) Lites, B. W.: 1973, *Solar Phys.* **32**, 283.
(14) Muller, E. and Mutschlecner, P.: 1964, *Astrophys. J. Suppl.* **9**, 1.
(15) Nussbaumer, H. and Swings, J. P.: 1970, *Astron. Astrophys.* **7**, 455.
(16) Payne, C. H.: 1925, *Stellar Atmospheres,* Harvard Observatory Monographs, Cambridge, Mass. no. 1.
(17) Richter, J. and Wulff, P.: 1970, *Astron. Astrophys.* **9**, 37.
(18) Rogerson, J. B.: 1969, *Astrophys. J.* **158**, 797.

(19) Ross, J. E.: 1970, *Nature* **225**, 610.
(20) Ross, J. E.: 1973, *Astrophys. J.* **180**, 599.
(21) Russell, H. N.: 1929, *Astrophys. J.* **70**, 11.
(22) Teplitskaya, R. B. and Vorob'eva, V. A.: 1964, *Soviet Astron.* **7**, 778.
(23) Unsöld, A.: 1948, *Z. Astrophys.* **24**, 306.
(24) Warner, B.: 1964, *Monthly Notices Roy. Astron. Soc.* **127**, 413.
(25) Warner, B.: 1968, *Monthly Notices Roy. Astron. Soc.* **138**, 229.
(26) Weidermann, V.: 1955, *Z. Astrophys.* **36**, 101.
(27) Withbroe, G. L.: 1969, *Solar Phys.* **9**, 19.

TABLE II

HD	NAME	SPECT.TYP	VM	MV	B-V	U-B	1-X	θ	l	LB B	LB PT	[M]	[y]	∑+	STANDARD	[Fe/H]	REF
26		B4VP	8.22*		1.08	0.53**		1.02	4			-0.5	-0.12		EPS VIR	-0.67	18
358	21ALF AND	B8P	2.02		-0.11	-0.47	-0.21	0.36	1	3.5				+5	SUN	-1.5	211
886	886AM PEG	B2IV	2.83		-0.23	-0.87	-0.39						+0.20		SUN	+0.48	187
1461		[G51V-V]	6.45	4.5	0.68	0.29		0.85	4					+5.1	SUN	+0.43	74
1461								0.97	3						HR 483	+0.23	128
1909		[B91V]	6.66	3.80	-0.03	-0.51**		0.46	1	4				+4	NORMAL A *	+0.08	160
1909								0.44	1	3.25					53 TAU	+1.04	171
2151	BET HYI	G2IV	2.79		0.62	0.11		0.94	2	4				+2.1Y	SUN	-0.30	20
2151								0.88	1	3.80				+1.5	SUN	-0.31	86
2421		[A2V]	5.17		0.04	-0.01**		0.53	1	4.0				+5	NORMAL A *	-0.1	145
2628	28 AND	AM	5.19		0.25	0.08**		0.73	2		+1.98			+4.8	NORMAL *	+0.76	143
2665		[G5III]	8		0.77			1.06	1	1.80	-1.51			+0.6	SUN	-1.56	42
3360	17 ZET CAS	B2V	3.61		-0.19	-0.89									SUN	+0.48	187
3443		G5V	5.56	5.7	0.71		1.42*	0.93	1	4.57				+1.5	SUN	-0.16	208
3546	30 EPS AND	G5III+	4.37		0.87	0.47		1.17	2	2.44	-0.75			+2.0	6AM TAU	-0.75	43
3627	31DEL AND	K3III	3.21		1.28	1.48									SUN	0.00	21
3651	54 PSC	K0V	5.84	5.75	0.85	0.57		0.95	1	4.68				+1.5	SUN	-0.06	61
3651								1.10	1		+1.75			+2.8	SUN	-0.32	126
3651								0.98	1	4.4				+1.9	SUN	-0.17	151
4614	24ETA CAS	G0V	3.45	4.60	0.58	0.02	0.61	0.89	1	4.37				+1.5	SUN	-0.17	61
4813	19PHI 2CET	F8IV	5.20	4.23	0.50	-0.01	-0.74	0.84	1			+0.21	-0.02		SUN	+0.03	19
5015		F8V	4.83	4.1	0.53	0.12	-0.78	0.85	1	4.13				+0.6	SUN	+0.06	62
5015								0.84	1	4.11			+0.05		SUN	+0.10	233
5544		K0III(P)	7.71*				0.16	1.20	1						IOT,KAP GEM,EPS VIR, ETA HER	+0.33	25
5737	ALF SCL	B4P	4.30		-0.18	-0.57		0.32	1	3.4				+2.5	* *	+0.7	205
5737								0.29	1	3.60				+1.5	* *	+0.9	206
5780		K5J1-111	7.61*		1.42	1.69**		1.53	1	1.20				+0.9	SUN	-0.43	152
6497		K2111+	6.43		1.18	1.27		1.12	1	2.0				+5.0	SUN	-0.40	41
6497								1.14	1	2.80				+1.5	EPS VIR	+0.02	165
6582	30MU CAS	G5VP	5.12	5.75	0.69	0.09		0.97	1	4.20				+5	SUN	-0.57	44
6582								0.96	1				+0.00		SUN	-0.62	57
6582								0.95	1	4.61		-0.51		+1.5	SUN	-0.71	61

TABLE II (CONTINUED)

HD	NAME	SPECT.TYP	VM	MV	B-V	U-B	I-X	θ	[1 LG θ/LG FE	[Fe]	[C]	Σ+	STANDARD	[Fe/H]	REF
6755		F8V	7.68 *		0.71	0.08 **		0.97 1	2.80 -0.43			+0.7	SUN	-1.04	42
6833		G8III	6.74 *		1.18	0.89 **		1.14 1	1.0			+2.4	SUN	-0.85	41
6833								1.16 1	1.50				SUN	-0.83	224
7383		A0IV-V	10.07 *		0.13 -0.35 **			0.63 2	+0.20			+9.5	A1+ CY6+HD 67456	-1.0	114
7383								0.56 3	0.87				SUN	-0.22	207
8890	1ALF UMI	F7+B1B-11	2.04 *	3.06	0.60	0.38	-0.80	0.89 2				+6.3	SUN	0.00	136
9826	50UPS AND	F8V	4.08 *		0.54	0.06	0.97	0.84 1	3.91	+0.08 +0.23		+1.3	SUN	-0.23	62
9826								0.83 1	+0.53				SUN	-0.11	75
9826								0.83 1	4.10				SUN	-0.14	98
9996		A0F	6.30 *	4.66	0.62	0.11	0.53	0.47 1	3.7			+4	SUN	+0.86	95
10307		G2V	4.94					0.85 4	4.38	-0.01			SUN	+0.20	3
10307								0.87 1			-0.01		SUN	-0.03	62
10307								0.84 4			+0.04	+0.7	SUN	+0.16	74
10307								0.84 4					SUN	+0.14	85
10380	30OMI	F8C K3III	4.43	5.72	1.06	1.56	2.01	1.26 1	1.50				SUN	0.00	21
10380												+1.5	SUN	-0.35	152
10700	52TAU CET	G8VP	3.50	5.91	0.72	0.21	1.28	0.77 2		-0.44			SUN	-0.39	5
10700								0.91 1		-0.03			SUN	-0.13	19
10700								0.95 1	4.59	-0.44		+1.5	SUN	-0.39	26
10700								0.74 1					SUN	-0.34	61
10780		K0V	5.59		0.81	0.40	-1.03	0.73 1	4.60			+1.3	SUN	+0.36	61
11007		[G4 0]	5.70 *		0.58	0.04 **		0.84 3	4.09			+0.7	SUN	-0.24	62
11397		G8III	8.95 *		0.69	0.10 **		0.89 1	5.00				SUN	+0.20	152
11909	81UPI	K1P	5.08		0.92	0.71 **		1.19 2	2.42 -0.58			+2.0	66AM TAU	-0.37	43
12929	13ALF ARI	K2III	2.00		1.15	1.18	1.18	1.12 1	2.4			+1.6	SUN	-0.21	151
13555	17ETA ARI	F5V	5.23	0.8	0.44	-0.03 **		0.82 1	3.70				UPS ANU	-0.30	98
13974	8BEL IKI	G0V	4.87	4.8	0.61	0.02		0.88 4		-0.28 -0.03			SUN	-0.43	3
13974								0.93 2		-0.54			SUN	-0.51	5
13974								0.90 2		-0.19 -0.03			SUN	-0.18	19
13974								0.87 2					SUN	-0.24	85
15144		A7P	5.84 *		0.14	0.09	-0.16	0.60 1	4.0			+7.59	UMI PEG	+0.03	190
16234	31	ARU [dF]	5.64 *		0.49	-0.03 **		0.85 1	3.95			+1.3	SUN	-0.49	62
16417	LAM 2FOR	G5IV	5.78 *	4.0	0.66			0.87 1	4.09	-0.19		+1.3	SUN	-0.20	86
16458		[K0F]	5.78 *					1.08 4		-1.57		+5.5	EPS VIR	-0.2	66
16458													P1 6 UKI	-0.18	188

TABLE II (CONTINUED)

HD	NAME	SPECT.TYP	VM	MV	B-V	U-B	1-X	θ	1	LB G/LB PE	[Fe]	[γ]	ξ+	STANDARD	[Fe/H]	REF
16895	13THE HER	F7V	4.12 *	3.62	0.49	0.90	1.03 *	0.86	2	4.12	-0.07	-0.01		SUN	+0.07	19
16895								0.83	1				+1.3	SUN	-0.26	62
17506	15ETA PER	K3IB+B9V	3.76		1.69	1.90	1.58	1.47	1	-3.2			10.3		-0.02	126
17948		[-4V]	5.56		0.44	-0.12 **		0.82	1	3.60				UPS AND	-0.29	98
18296	21 HER	A0P	5.10		-0.01	-0.24 **		0.45	1	3.0			+4.79	UM1 PEG	-0.12	190
18474	34P	G4P	5.52		0.89	0.61 **		1.06	1					IOT,KAP GEM,EPS VIR, ETA HER	+0.15	25
18474								1.05	1	2.67				SUN	-0.20	121
18769	4Y AR1	AM	5.86		0.14	0.15 **	1.01 *	0.59	2				+4.8	2 HYA	+0.80	228
19373	10T HER	G0V	4.04	3.7	0.60	0.12		0.85	4	4.09	-0.15	+0.05		SUN	+0.14	3
19373								0.87	2		+0.02	+0.02		SUN	+0.26	19
19373								0.85	1				+1.3	SUN	-0.01	62
19373								0.84	4		-0.01	+0.09		SUN	+0.05	75
19373								0.87	4	+0.44				SUN	+0.01	85
19445		G5 VI	8.04 *		0.46	-0.24 **		0.8	1	4.80	-0.25			SUN	-1.75	3
19445								0.88	2	+0.90				SUN	-0.77	11
19445								0.86	4	4.00				SUN	-1.75	38
19445													1-2	SUN	-2.55	58
19476	22KAP PER	K0III	3.81		0.99	0.83 **	1.52 *	1.02	1	3.50			+1.51	SUN	+0.08	152
20630	96KAP CET	G5V	4.82	4.99	0.68	0.19	0.93	0.85	2	4.40	+0.43	+0.12		SUN	+0.38	19
20630								0.89	1	4.40			+1.6	SUN	+0.08	62
20766	ZE1 1RET	G2V	5.53	5.28	0.64	-0.64	0.88	0.91	1	3.80			+2.02		-0.37	45
20766								0.92	1	-0.50			+0.5	SUN	-0.10	152
20794	82 ERI	G5V	4.26	5.29	0.71	0.22	1.02	0.94	1	4.40				SUN	-0.34	63
20807	ZE2 2RET	G1V	5.23	4.98	0.60	0.60	0.93	0.88	1	4.00	-0.04			SUN	-0.04	152
20807								0.90	1	4.51			+1.3	SUN	-0.32	208
20902	33ALF PER	F5IB	1.79		0.48	0.59	0.45	0.82	1	1.7	+0.1		+0.5	SUN	+0.05	156
20902								0.80	1	+1.16			+1.3	SUN	-0.3	167
21389		A01A	4.55 *		0.56	-0.11	1.20	0.50	1	1.06			+9.4	ALF CY6	-0.40	172
22049	18EPS ERI	K2V	3.73 *	6.13	0.88	0.58	1.35	0.99	1	4.61	+0.22	-0.03	+1.3	SUN	-0.31	61
22049								1.01	1	4.4			+2.3	SUN	-0.19	151
22484	10 TAU	F8V	4.29	3.21	0.57	0.08	1.04	0.85	1	3.87			+1.3	SUN	+0.37	19
22484								0.85	1	3.96	-0.14		+1.1	SUN	-0.16	61
22484								0.84	1					SUN	-0.12	86
22615		A4III	6.51		0.15	0.19 **		0.59	1				10.0	SUN	+0.1	70
22879		F9V	6.68 *	4.6	0.54	-0.09 **		0.89	4	4	-0.48	-0.05		SUN	-0.57	3

TABLE II (CONTINUED)

HD	NAME	SPECT.TYP	VR	MV	B-V	U-B	I-X	θ	I	LG G	LG PE [Fe]	[S]	[γ]	ξ+	STANDARD	[Fe/H]	REF
23194		A5V	8.05*		0.20	**		0.59	1	4.0				+9.0	SUN	-0.2	70
23249	23UEL	K0IV	3.55*	3.77	0.92	0.67	1.34*	1.04	2					+1.9	SUN	-0.09	5
23249								1.24	2		-0.85				E1A CEP	+0.79	13
23249								0.87	4		-0.8				SUN	-0.05	22
23249								1.05	1	3.36	-0.86				SUN	-0.27	86
23269		[A2M]	5.43		0.09	0.12	0.14	0.85	1	4.3	-0.63			+2.0	SUN	-0.03	23
23277	P							0.55	1		+1.17	+0.20			SUN	+0.52	154
23277	S							0.55	1		+1.08	+0.18			SUN	+0.80	154
23386		G6V	9.8					0.88	1	4.3				+1.75	SUN	-0.07	23
23387		A1V	6.77		0.16	0.08	**	0.54	1	4.0				+2.5	SUN	-0.3	70
23464		G0V	8.66		0.60	0.10	**	0.86	1	4.3				+2.5	SUN	+0.08	23
23607		A7V	8		0.26	0.10	**	0.64	1	4.0				+9.5	SUN	-0.3	70
23631		A2V	6.30		0.01	-0.05	**	0.53	1	4.0				+4.5	SUN	-0.1	70
23713		F6V	9.25		0.55	0.12	**	0.81	1	4.3				+2.0	SUN	+0.03	23
23732		F4V	9.13		0.47	0.07	**	0.76	1	4.3				+1.75	SUN	+0.11	23
23924		A7V	9		0.22	0.13	**	0.63	1	4.0				+9.5	SUN	-0.3	70
23964		A0V	7		0.06	-0.06	**	0.50	1	4.0				+4.0	SUN	-0.1	70
24368		[A0]	7.16	7.10	0.12	0.10	**	0.53	1	4.0	-0.4 / -0.47	-0.92	+0.00	+5.0	SUN	+0.2	70
25329		K1V	8.52		0.88	-0.48	**	1.2	2					+3.30	SUN	-2.30	27
25329								1.03	1					+3.5	UMI PEG	-1.32	46
25823	41	TAU A0F	5.19		-0.14	-0.48	**	0.34	1	3.5				+3.5	SUN	-0.15	190
25825		[G0]	7.85		0.59	0.10	**	0.86	1	4.3						+0.12	23
26462	45	TAU F4V	5.73		0.36	0.00	0.19	0.63	1	4.44	-0.15				HD 28344	+0.1	223
26630	51MU	PE0 G0IB	4.13		0.95	0.84	1.54*	0.99	1	1.30				+3.6	SUN	+0.07	132
26736		[G5]	8.0				**	0.88	1	4.3				+2.7	SUN	+0.19	23
26756		[G5]	8.46		0.70	0.24	**	0.89	1	4.3				+3.4	SUN	+0.12	23
26756								0.93	1	4.44	-0.04				HD 28344	-0.15	223
26767		[G0]	8.06	5.99	0.64	0.17	**	0.87	1	4.3			-0.22	+4.1	SUN	+0.23	23
26965	400M1 2ERI	K1V	4.42		0.82	0.44	1.30*	0.99	2			+2.30		+2.4	SUN	-0.19	5
26965								1.17	3						ALF LYR	+0.01	126
27295	53	TAU B9VP*	5.28		-0.07	-0.26	**	0.46	1			+2.75		+3	SUN	-0.57	160
27295								0.45	2						ALF LYR	-0.86	179
27295								0.41		3.5					SUN	-0.21	214
27295								0.41	1	4.0		+2.75		+2.40	UMI PEG	-0.71	190
27295								0.45	2						NORMAL A *	-0.65	183

TABLE II (CONTINUED)

HD	NAME	SPECTYPE	VM	MV	B-V	U-B	1-X	θ	l	LG G	LG FE	[FE]	[V]	ξ+	STANDARD	[Fe/H]	REF
27371	54GAM	TAU K0III	5.66		0.99	0.81	1.45 *	0.99	2			-0.99	+0.02		SUN	+0.10	6
27371								1.01	1	2.5				+4.0	SUN	0.00	21
27371								1.15	2						SUN	-0.08	23
27371								1.01	1	2.5	+2.24			+4	ALF BOO	+0.61	47
27371								1.03	1	2.7				+1.4	SUN	-0.1	138
27371								1.02	1	2.70				+1.7	SUN	+0.05	151
27371								1.00	1	2.70			+0.01		SUN	+0.24	152
27371															SUN	-0.12	224
27383		+9V	5.88	*	0.56	0.09		0.84	1	4.3				+2.5	SUN	+0.23	23
27561		+5V	5.61	*	0.41	0.00		0.75	1	4.44			-0.10		HD 28344	0.00	223
2757O		[66]	9.10	*	0.65	0.14		0.89	1	4.3				+3.2	SUN	+0.24	23
27628 60		TAU AM	5.72		0.32	0.10	0.48	0.67	1	4.0			+0.20	+7.8	HR 114,906,2085,4825	0.00	123
27628								0.67	1	3.86			+0.20		45 TAU	+0.16	235
27628								0.67	1	2.86					45 TAU	+0.14	235
27697 61DEL		TAU K0III	5.76		0.99	0.82	1.40 *	1.03	2			-1.04	-0.01		SUN	+0.10	6
27697								1.01	1	2.5				+4.0	SUN	0.00	21
27697								1.01	1	2.5				+4	SUN	-0.13	23
27697								0.99	1	3.00				+1.46	SUN	-0.1	138
27697								1.00	1	2.70			0.00		SUN	+0.07	152
27697															SUN	-0.01	224
2749 63		TAU AM	5.84		0.30	0.13	0.43	0.65	1	4.3				+5.0	SUN	+0.65	65
2749								0.65	1	4.0				+7.0	SUN	+0.81	88
2749								0.61	1		+1.57		+0.25	+4.3	HR 114,906,2085,4825	+0.39	111
2749								0.60	1	4.4					HOKMAL *	+0.28	123
2749								0.61	1	4.4					SUN	+0.14	143
2749								0.60	1	3.80					SUN	+0.57	156
2749								0.66	1	2.80			+0.25		SUN	+0.57	169
2749													+0.25		45 TAU	+0.37	235
2749															45 TAU	+0.36	235
27808		[+5]	8.4					0.82	1	4.44				+4.0	HD 28344	-0.25	223
27819 64		TAU A7V	4.80	*	0.15	0.13	0.23	0.62	1	4.00			0.00	+5.6	SUN	+0.17	65
27819															HD 28344	-0.2	223
27836		[60]	7.62	*	0.60	0.12		0.88	1	4.3				+5.5	SUN	+0.16	23
2765		62V	7.80	*	0.60	0.13		0.88	1	4.3				+3.2	SUN	+0.12	23
27962 68		TAU A2IV	4.30		0.04	0.06	-0.10	0.55	1	4.0			+0.15	+5.0	HR 114,906,2085,4825	+0.70	65
27962								0.54	1	4.0				+3.2	HD 28344	+0.55	123
27962								0.62	2						HD 28344	0.00	223
27962															2 HYA	+0.89	228
27989		[60]	7.56	*	0.68	0.24		0.91	1	4.3			+0.14	+4.0	SUN	+0.08	23
28068		61V	8.06	*	0.63	0.17		0.84	4	4.3			+0.10		SUN	-0.04	3
28068								0.84	4	4.3			+0.14		SUN	+0.04	23
28068								0.84	4	4.3					SUN	+0.00	28
28068								0.96	3						SUN	+0.05	29
28068								0.87	4	4.50					SUN	+0.07	30
28068								0.86	1	4.50			+0.09		SUN	+0.23	85
28068															SUN	+0.05	152
28068															SUN	+0.09	224

TABLE II (CONTINUED)

HD	NAME	SPECT.TYP	VM	PV	B-V	U-B	I-X	D	n	⁽¹⁾L₆ b⁽¹⁾L₆ Pt	[Pt]	[V]	N)+	STANDARD	[Fe/H]	Ref
280999		68V	8.12	*	0.86	0.20		0.87	4		+0.20	+0.08		SUN	+0.08	3
280999			8.4	*				1.00	3			+0.08	+4.0	SUN	+0.16	30
28304		[88]					* 1.50	1.01	1	2.5				SUN	+0.09	23
28305/ALF5	TAU K0III		3.54	*	1.01	0.87		1.02	1	2.5	-0.93	+0.10		SUN	+0.10	6
28305						**		1.01	1	2.5			*	SUN	+0.1	138
28305								1.03	1	2.7			+1.8	SUN	+0.07	151
28305								1.04	1	3.00			+1.59	SUN	+0.19	152
28305								1.04	1	2.70		+0.08		SUN	+0.12	224
28307//	TAU K0III		3.85	*	0.95	0.71	1.41	1.00	2		+2.43	+0.10		SUN	+0.03	6
28307								1.14	2	2.7			+1.4	ALF BUU	+0.68	47
28307								1.01	1	2.7				SUN	+0.18	151
28307								1.04	1	2.70		+0.02		SUN	+0.06	224
28344		62V	7.85	*	0.61	0.13		0.82	4		+0.18	+0.09		SUN	-0.06	3
28344						**		0.82	4		+0.25	+0.10		SUN	+0.13	28
28344								0.95	3		+0.20	+0.10		SUN	+0.18	29
28344								0.93	3			+0.09		SUN	+0.07	30
28344								0.86	1	4.50				SUN	+0.36	85
28344								0.87	1	4.50		+0.09		SUN	+0.05	152
28344														SUN	+0.02	224
28483		[45]	7.57				* -0.58	0.79	1	4.44		+0.09	-0.05	HD 28344	-0.1	223
28546/81	TAU AM		5.48		0.26	0.10		0.63	1	4.0			+6.0	HK 114,906,2085,4825	+0.14	123
28546						**		0.62	1	3.87		+0.10		45 IAU	+0.34	235
28992		[8]	7.94	*	0.64	0.15	* 1.50	0.87	1	4.3			+5.0	SUN	+0.19	23
29139/87ALF	TAU K5III		0.86	-0.6	1.54	1.92		1.35	1	1.20			+1.53	SUN	0.00	21
29139								1.27	1	1.3			+1.9	SUN	-0.10	152
29139														ALF BUU	+0.42	173
29140/88	TAU AM		4.26		0.18	0.12	-0.29	0.60	1	4.0	+1.23	+0.36		HK 114,906,2085,4825	+0.27	77
29140						**		0.60	1	4.0			+6.0	SUN	+0.01	123
29310		[90]	7.54	*	0.19	0.12		0.88	1	4.3			+3.5	SUN	+0.05	23
30210		AM	5.37		0.62	0.09	-0.26	0.60	1	4.0		+0.40	+8.5	HK 114,906,2085,4825	+0.31	123
30210						**		0.57	1	3.71				45 IAU	+0.56	235
30455		62V	6.97	*	0.62	0.20		0.85	4		-0.01	-0.13		SUN	-0.26	3
30455								0.97	3		+0.17	-0.13		SUN	-0.09	29
30562		[60V]	5.76	4.2	0.62	0.14		0.86	1	3.75	-0.17		+1.3	SUN	+0.13	86
30649		G1V-VI	6.98		0.58			0.85	4		-0.08	-0.02		SUN	-0.20	3
30649								0.87	2		-0.21	-0.02		SUN	-0.32	51
30652/1PI ZORI	F6V		3.18	3.78	0.45	-0.01	* 0.78	0.78	4		+0.67 +0.04	-0.04		SUN	+0.16	64
30652								0.78	4		+0.22	+0.20		SUN	+0.18	75

TABLE II (CONTINUED)

HD	NAME	SPECT.TYPE	VM	MV	B-V	U-B	I-X	θ	lLoG G PE	[PE]	[Y]	ξ+	STANDARD	[Fe/H]	KEY
30676		[F8]	7.12 *		0.56	0.06 **		0.84 1 4.3				+5.0	SUN	+0.17	23
30810		F6V	6.76 *		0.54	0.05 **		0.84 1 4.3				+5.0	SUN	+0.16	23
31295	7P1 1UKI	A0V	4.66		0.08	0.09	0.14	0.59 1 4.0	+2.0			+4.0	ALF LYR	-0.5	1/4
31398	3101 AUR	K3II	2.66		1.53	1.78	1.44							0.00	21
32034		B9IA	10.01 *		0.10			0.61 2 1.29	+0.8				ETA LEU+13 MON	-0.8	213
32147/		[K3V]	6.21	6.40	1.06	1.00	1.34	1.06 1 4.5				+2.0	SUN	+0.02	150
32147/								1.06 1 4.4				+2.1	SUN	+0.02	151
32537	Y AUR	F0V	5.00	3.4	0.34	-0.01	-0.51	0.70 1 4.44				+5.7	SUN	+0.10	88
32549	11 UKI	AUF	4.68		-0.05	-0.10	-0.02					+4.79	UKI PEG	-0.34	190
32923	104 1AU	G4V	4.91	4.5	0.64	0.14 **		0.88 1 3.98				+0.7	SUN	-0.20	86
33254	16 UKI	AM	5.43		0.24	0.15	-0.52	0.62 1 4.0		+0.20	+7.2	HR 114,906,2085,4825	+0.4/	123
33254								0.62 1 3.78		+0.20		45 1AU	+0.63	235
33254								0.62 1 2.78				45 1AU	+0.64	235
33256	68 ERI	F5V	5.11	3.5	0.44	-0.05		0.82 1 4.18				+1.3	SUN	-0.60	61
33579		A31A-0(E)	9.5 *		0.18		**	0.65 2		+0.20		SUN	-0.31	103
33579								0.70 2 0.98	0.0		6-26	SUN	-0.2	129
33579								0.62 3 0.7	+0.25			SUN	-0.1/	185
34255		[CK4]	5.67 *		1.75	2.00 **	2.46	1.68 3	-2.05		+9.8	SUN	-0.23	126
34334	16 AUR	K3III	4.54 *		1.27	1.27	2.02	1.22 1 2.1			+1.6	SUN	-0.39	151
34411	15LAM AUR	G0V	4.74 *	3.84	0.62	0.13	0.57	0.85 4		+0.07 -0.04		SUN	+0.22	3
34411								0.97 3		+0.16 -0.04		SUN	+0.14	29
34411								0.86 1 4.11			+1.3	SUN	+0.35	61
34411								0.85 4				SUN	+0.07	85
34452		B9P	5.38		-0.19	-0.60 **	0.40	0.28 1 4.2			+8	SOLAR SYSTEM	+1.7	157
34816	6LAM LEP	B0.51V	4.28 *		-0.25	-1.01		0.16 1 4.05			+4.5	SUN	+1.0	105
35620	24PHI AUR	K3P	5.07		1.40	1.66	2.06	1.20 1 11.0		+0.10	+5.7	EPS VIR	-0.42	41
35620								1.22 1 11.90				GAM,DEL,EPS TAU,EPS,VIR	-0.11	99
35620								1.12 1 2.00		+0.11	+6.5	SUN	-0.4	138
35620								1.20 1 11.50				SUN	-0.09	224
36916		[B8111P]	6.73		-0.10	-0.58 **		0.35 1 3.8	+2.4			GAM GEM	+0.70	149
37043	101 UKI	B 2						0.28 1 4.0				HD 35912 Add B	+0.1	143
37058		N3VP	7.34 *		-0.16	-0.75 **	-0.47	0.28 1 3.5	+2.40		+4	SUN	+0.80	81

TABLE II (CONTINUED)

HD	NAME	SPECT.TYP	VM	MV	B-V	U-B	1-X	θ	1 LG G LG M	[Fe]	[y]	ξ+	STANDARD	[Fe/H]	REF
37160	40FHI 2GRI	G8III-IV	4.09		0.95	0.66	1.59 *						SUN	-0.30	21
37160												IOT,KAP GEM,EPS VIR,ETA HER	-0.18	25
37160											+1.86	6AM TAU HER	-0.73	43
37160											+9.5	SUN	+1.40	184
37763	6AM MEN	K4III	5.18		1.13	1.18		1.00 1 3.30				SUN	+0.40	152
38393	136AM	LEP F6V+K2V	3.60	4.05	0.47	0.01	0.76 *	0.89 2			+1.3	SUN	-0.07	19
38751	32	TAU G8III	4.87		1.01	0.79 **	1.68 *	1.27 2 2.36 -0.60	-0.40 +0.00		+1.93	6AM TAU	-0.11	43
39003	32MU	AUR K0III	3.97		1.14	1.09	1.72 *	1.09 1 2.5			+1.4	SUN	-0.10	151
39091	PI	MEN G3IV	5.64		0.60	0.11		0.89 1 3.94			+1.3	SUN	0.00	208
39364	150EL	LEP G8III	3.77		0.98	0.71	1.55 *	1.15 1				IOT,KAP GEM,EPS VIR, ETA HER	-0.36	25
39587	54CHI 1URI	G0V	4.41	4.43	0.59	0.08	1.04 *	0.87 2 4.44	+0.03 +0.05		+1.3	SUN	+0.25	19
39587								0.84 1	+0.06			SUN	-0.12	86
39853		K3+III-	5.66		1.53	1.84	2.38 *					SUN	0.00	21
39866		A21B	6.39		0.30	0.26 **		0.5 1 1.95 +1.44			+5.3	ALF CYG	+0.01	172
40183	34BEI S	AUR A2V	1.90		0.03	0.05		0.57 1 3.7			+8	NORMAL A *	0.00	176
40183								0.57 1 3.7			+8	NORMAL A *	-0.03	176
40932	61MU	URI AM	4.12		0.16	0.11	-0.27	0.89 3			+5.3	URI PEG	+0.76	190
41312		[5K3]	5.05		1.34		*	1.26 1 0.90			+1.0	SUN	-0.60	152
41357	40 S	AUR [AAM]	5.27		0.23	0.11	**	0.63 1	+1.01 +0.37			SUN	+0.55	154
41357								0.60 1	+1.14 +0.34			SUN	+0.80	154
43039	44KAP	AUR G8III	4.34		1.01	0.81	1.60 *	1.25 2 2.24 -0.72			+1.90	GAMMA TAU	-0.42	43
43039								1.01 1 2.28			+3.5	GAM,DEL,EPS TAU,EPS, VIR	0.00	138
44033		K31B	5.81		1.59	1.74	2.52 *	1.53 3 -2.70			10.0	SUN	-0.07	126
44537	46PS1 1AUR	K5IAB	5.02		1.97	2.2	2.55 **	1.65 3 -3.5			10.7	SUN	+0.08	126
44691	KK	LYN AM	5.64		0.24	0.12		0.62 1 4.0			+7.0	HK 114,906,2085,4825	+0.35	123
45348	ALF	CAR F0IB	-0.73		0.15	-0.15	0.12	0.73 2 +0.30			+4.95	SUN	-0.15	103
45829		K0IAB	6.63		1.58	1.67 **		1.33 3 -2.80			+9.4	SUN	-0.16	126
46300	13	MON A0IB	4.48		0.00	-0.17		0.63 2 +0.3			+3.6	SUN	-0.3	115
46407		[K0P]	6.26		1.11	0.78		1.20 2			+4.3	KAP GEM	-0.08	15
46407								0.99 2 -0.78			THE CEN,ALF IND	+0.08	16
47105	246AM GEM	A0IV	1.93		0.00	0.05	-0.03	0.53 1 3.7			+2.5	SUN	-0.05	65
47105								0.47 1 4.0				NORMAL A *	0.00	145
47105											+4.79	URI PEG	+0.05	190

TABLE II (CONTINUED)

HD	NAME	SPECT.TYP	VM	MV	B-V	U-B		I-X	Θ	l	Lb 6 Lb Ht	[Ht]	[y]	M+	STANDARD	[Fe/H]	KEF
47152	53 AUR	A0p	5.51		-0.01	-0.07	**	1.19	0.50	1	4.0			0.00	ETA LEO+ALF LYR	+0.4	134
48329	27EPS GEM	B8Ib	2.98		1.40	1.46			1.18	1	1.00	-1.95		+8.7	SUN	-0.09	126
48329									1.10	1				+5.3	SUN	-0.04	132
48682	56PSI 5AUR	G0V	5.22	4.37	0.56	0.06	**		0.88	2	2.5	-0.04	+0.00	+5.4	SUN	+0.15	19
48781	57PSI 6AUR	K1III	5.15	1.42	1.11	1.04	**	1.81	1.08	1	2.5				SUN	-0.10	41
48915	9ALF CMA	A1V	-1.47		0.00	-0.06	**		0.48	1	3.5			+5	SUN	+0.4	94
48915									0.56	1	4.0				SUN	+0.71	97
48915									0.52	1	4.3			+2.0	SUN	+0.87	107
48915									0.50	1	4.44			-0.36	SUN	-0.36	109
48915									0.49	1	4.2			+3.0	SUN	+0.8	117
48915									0.54	1	4.30			+3.0	SUN	+0.82	122
48915									0.52	1	4.3			+2.0	SUN	+0.70	141
48915															SUN	+0.81	159
50778	14THE CMA	K3+III	4.06		1.43	1.69		2.14	1.33	1	1.7			+1.9	SUN	-0.22	151
52005	41 GEM	[CK4]	5.62		1.62	1.86	**	2.43	1.60	3		-2.20		+8.5	SUN	-0.29	126
52711		G8IV	5.85		0.59	0.07	**		0.86	1	4.4			+0.8	SUN	-0.15	61
53244	23GAM CMA	B8II	4.10		-0.11	-0.48		0.11	0.79	2	+0.56			+1.51	UMI Pt&G	-0.24	190
54605	25 DEL CMA	F8Ia	1.84		0.67	0.54		0.58	0.85	4	3.98			12	SUN	+0.01	153
55575		G0V	5.58		0.58	0.03			0.85	4					SUN	-0.21	3
55575									0.87	1		-0.09	-0.04	+0.4	SUN	-0.44	61
58207	60I01 GEM	G9III+	3.80		1.04	0.84		1.5	1.14	1	3.98			+5	SUN	+0.16	106
58343		B3V	5.33		-0.05	-0.60	**		0.25	3	3.75			+5	SUN	+0.89	106
60178	66ALF GEM	A1V	2.85	2.04	0.04	0.02			0.57	1	4.25			+3.0	ALF LYR	+0.45	124
60179	66ALF GEM	Am	1.99	1.14	0.03	0.01	**		0.49	1	4.0			+3.0	ALF LYR	+0.98	124
61421	10ALF CMI	F5IV-V	0.34	2.64	0.42	0.03		0.36	0.80	4	-0.16	-0.20			SUN	-0.29	48
61421									0.76	4	+0.85	-0.40	+0.1/	+2./	SUN	+0.03	64
61421									0.77	1	+0.1/		+0.07		SUN	+0.07	75
61421									0.86	2	+0.5				SUN	0.00	84
61421									0.75	1	+0.96				SUN	+0.22	156
61421															SUN	-0.4	167
62345	77KAP GEM	G8III	3.57		0.92	0.70		1.38	1.29	3		-0.95		+4.1	SUN	0.00	15
62345									1.14	1					SUN	-0.20	25
62509	78BE1 GEM	K0III	1.15		1.00	0.86		0.98	1.25	3		-0.95		+1.3	SUN	-0.51	126
62644		G5IV	5.05	3.8	0.78				0.98	1	3.12	-0.49		+1.3	SUN	-0.35	86
63077		G0V	5.36	4.36	0.58	0.03			0.90	2	3.70			+0.6	SUN	-0.98	32
63077									0.95	4	4.15				SUN	-0.8	71
63077									0.89	1		+1.3			SUN	-0.82	86
63077									0.87	1	4.15			+0.7	SUN	-0.90	191
63077									0.89	1	4.15			+1.3	SUN	-0.70	208

TABLE II (CONTINUED)

HD	NAME	SPECT.TYP	VM	AV	B-V	U-B	I-X	θ	n	LG g	Pt	[Pt]	[y]	ξ+	STANDARD	[Fe/H]	Ref
64491		[A3IV]	6.14		0.28	-0.02	**	0.73	2		+1.8			+0.7	NRU BEM	+0.2	32
67456		Am	5.37		0.10			0.65	2					+5.9	SUN	-0.06	114
67523	15RHO PUP	F6II	2.88		0.43	0.19	0.73*	0.83	2	2.5	+0.4			+1.3	SUN	+0.13	156
69897	18CM1 CNC	F6V	5.13	4.1	0.47	-0.06		0.82	1	4.32					SUN	-0.52	61
71369	10M1 UMA	G4II-III	3.36		0.85	0.52	0.81	0.97	2	2.30				+0.83	SUN	-0.02	152
72324	32UPS 2CNC	G9III	6.36		1.02	0.88	2.14*	1.1/	1						IOT,KAP GEM,EPS VIR, ETA HER	+0.32	25
72905	3P1 1UMA	G0V	5.64	4.67	0.62	0.07		0.87	1	4.40				+1.3	SUN	-0.27	61
72988	3 HYA	[A2P]	5.72		-0.03	-0.01		0.45	1	4.0					SUN	+1.6	163
73665	39 CNC	[G9III]	6.39		0.78	0.83	0.97	1.01	1			+0.08	+0.07		6AM TAU	-0.04	12
73666		A1V	6.61*		0.01	0.02	**	0.53	1	4.0				+4.0	SUN	+0.64	65
73666								0.52	1	4.00			+0.05		SUN	+0.1	70
73666															HD 28344	-0.5	223
73710		K0III	6.44*		1.02	0.90	**	1.03	1			+0.09	-0.02		6AM TAU	-0.17	12
74521	49 CNC	A0P	5.86		-0.11	-0.24	**	0.89	3					+5.02	UM1 PEG	+0.65	190
75333	14 HYA	[B9P]	5.31		-0.10	-0.35								+1.91	UM1 PEG	-0.27	190
75732	55RHO 1CNC	[G8V]	5.94	5.3	0.86	0.64	**	1.13	1	4.5	+1.65			+5.3	SUN	-0.15	126
75732								0.97	1	4.4				+1.1	SUN	+0.24	150
75732								0.97	1					+1.9	SUN	+0.11	151
76294	16ZE1 HYA	K0III	3.12		1.00	0.82	1.03	1.13	3		+0.05			+5.6	SUN	-0.12	126
78209	15 UMA	AM	4.46*		0.27	0.12	-0.58	0.71	1	4.20				+5.0	SUN	+0.24	91
78316	76KAP CNC	B8IIIP	5.24*		-0.11	-0.43	**	0.36	1	3.6				+5	SUN	+0.39	202
78316								0.35	1	3.5				+1.58	UM1 PEG	-0.25	190
78362	14TAU UMA	AM	4.65		0.35	0.15	**	0.6/	1	4.0	0.0			+6.5	HR 114,996,2085,4825	+0.52	125
78362								0.86	2	4.0	0.0				SUN	-0.07	156
78362								0.86	2		+0.12			+5.8	SUN	+0.16	230
79452		G6III	5.97		0.86	0.37		1.21	2	2.20	-0.88			+1.89	6AM TAU	-0.85	43
79469	22THE HYA	B9.5V	3.88		-0.07	-0.12		0.45	1	4.2	+2.7			+2.1	ALF LYR	+0.4	174
82210	24 UMA	G4III-IV	4.58	2.9	0.77	0.33	1.31	0.95	1	3.58				+1.3	SUN	-0.38	61
82328	25THE UMA	F6III-IV	3.18	2.0	0.46	0.03	0.45	0.80	4	4.0	+0.36	-0.09	+0.10	SUN	-0.03	75
82328								0.87	2	3.5	+0.6				SUN	+0.01	156
82885	11 LMI	G8IV-V	5.41	5.60	0.77	0.44	0.99	0.92	1	4.61				+1.3	SUN	0.00	61

TABLE II (CONTINUED)

HD	NAME	SPECT.TYP	VM	MV	B-V	U-B	1-X	Θ		[FE]	[γ]	Σ+	STANDARD	[FE/H]	REF
83548	[RU]		5.49		1.00	0.67		0.92 2		-0.29		+4.5?	THE LEO,ALF TAU	+0.14	16
83548								0.99 1		-0.16		+2.7	ALF BOO	+0.5	17
84441	17EPS LEO	G0II	2.96	*	0.81	0.46	1.19	0.93 1	2.4			+2.6	SUN	-0.13	78
84737		G2V	5.11	*	0.62	0.08	0.51	0.86 1	4.27			+1.3	SUN	-0.04	61
85503	24MU LEO	K2III	3.90	*	1.22	1.40	1.73	1.13 1	2.20		+0.06	+.5	EPS VIR	-0.08	99
85503								1.06 1	2.7				GAM,DEL,EPS TAU,EPS,VIR	+0.1	138
85503								1.13 1	2.4			+1.8	SUN	-0.01	151
85503								1.14 1	2.3			+1.5	SUN	+0.03	217
85503								1.14 1	2.3			+2.0	SUN	-0.11	217
85504	SEX A1V		6.02		-0.04	-0.09		0.50 1		+0.15	-0.03		ALF CMA	+0.20	141
86728	20 LMI G4V		5.36	4.6	0.66	0.28		0.84 4		+0.51	+0.07		SUN	+0.34	3
86986								0.84 4		+0.53	+0.07		SUN	+0.34	31
86986		A1V	7.99	*				0.88 1	4.12			+1.3	SUN	-0.08	61
87737	30ETA LEO	A0Ib	3.48		-0.04	-0.21	-0.01	0.66 1	2.8	+1.20		+4.6	SUN	-0.3	115
87737								0.63 2	2.05	+0.4 +1.31		+2.25	SUN	+0.10	116
88218		[G0]	6.12	4.5	0.60	0.16		0.49 5			-0.65	+1.3	SUN	-0.42	86
88284	41LAM HYA	K0III	3.62		1.00	0.92	1.46	0.91 1	3.36			+1.95	GAM TAU	+0.05	43
88284								1.21 2	2.25	-0.37		+4	GAM,DEL,EPS TAU,EPS,VIR	+0.2	138
89125	39	LEO[F8V]	5.82	4.7	0.50	-0.05	0.63	1.01 1	2.9			+1.3	SUN	-0.19	61
89388		K5III	3.44		1.55	1.70		0.84 1	4.51				ALF BOO	+0.3	177
89484	41GAM 1LEO	K0III	2.61		1.15	1.00	1.19	1.24 1					IOT,KAP,GEM,EPS,VIR,ETA HER	+0.10	25
89484								1.20 1	2.27				SUN	-0.22	34
89484								1.17 1	2.9	-1.04		+1.91	GAM TAU	-0.49	43
89485	41GAM 2LEO	G7III+	3.80		0.91	0.65		1.32 1	1.79				IOT,KAP,GEM EPS,VIR,ETA HER	-0.26	25
89485								1.09 1	2.14	-0.86		+1.93	GAM TAU	-0.71	43
89822	P	A0P	4.93		-0.07	-0.15	0.02	1.24 2		+2.44		0.0	ALF LYR	+0.28	180
89822	S							0.87 4		+2.44			ALF LYR	+0.51	180
90277	30	LMI F0V	4.73		0.25	0.18	-0.40	0.91 4		+1.24	-0.06 +0.00	+6.03	MURHAL *	+0.20	143
90508		G1V	6.44	5.0	0.60	0.05		0.75 2	3.0				SUN	-0.23	3
90508								0.85 4					SUN	-0.23	85
90537	31RET LMI	G8III-IV	4.20	4.44	0.90	0.65	1.55	0.88 4	4.41	-0.05	+1.51	EPS VIR	+0.21	139
90839	36	UMA F8V	4.82		0.52	-0.01	0.76	0.99 1				+1.3	SUN	-0.23	62
91324		[F6V]	4.88	3.5	0.50			0.83 1	3.90			+1.7	SUN	-0.60	191

TABLE II (CONTINUED)

HD	NAME	SPECT.TYP	VM	MV	B-V	U-B	1-X	θ	[1]	LG G	LG FE	[FE]	[Y]	ξ+	STANDARD	[FE/H]	KEF
92826		[K2P]	7.08	*	1.36	**		0.98			-0.44			+4.27	THE GEN+ALF IND	+0.08	16
94247	44 UMA	K3III	5.20		1.36	1.51	2.05	1.21	1	2.30				+1.54	SUN	0.00	152
94264	46 LMI	K0III-IV	3.82		1.05	0.91	1.58	1.07	1	2.5			0.00	+2.5	SUN	-0.16	41
94264								1.06	1	3.0				+1.85	EPS VIR	-0.28	139
95128	47 UMA	B0V	5.06	4.4	0.61	0.13		0.86	1	4.31				+1.5	SUN	-0.02	82
95272	7ALF UKT	K0III	4.08		1.09	0.98	1.51	1.25	2	2.46	+0.54			+1.93	6AM TAU	-0.12	43
95418	488E1 UMA	A1V	2.36		-0.02	0.00	0.21	0.47	1	4.39				+5.0	SUN	+0.78	122
95689	50ALF UMA	K0II-III+F7V	1.79		1.07	0.92	1.05	1.14	4	2.03	-0.71			+1.89	SUN	-0.19	34
95689								1.27	2						6AM TAU	-0.23	43
96446		B2IIIP	6.68	*	-0.16	-0.87	**	0.20	1	4.20				+5	H *	-0.6	203
96833	52PSI UMA	K1III	3.01		1.14	1.12	1.63	1.23	2		-1.01			+2.94	SUN	-0.39	49
97633	701HE LEO	A2V	3.31		-0.02	0.08	0.03	0.52	1	3.5				+4.0	ALF LYR	+0.58	124
97633								0.65	2						SUN	+0.4	133
97907	73 LEO	K3III	5.34		1.20	1.10	1.95	1.32	2	2.07	-0.85			+1.95	6AM TAU	-0.17	43
98230	53X1 UMA	B0V	4.87	4.90	0.59	0.03	**	0.90	2						SUN	-0.12	19
98231	53X1 UMA	B0V	4.41					0.86	2			+0.12	-0.06		SUN	-0.01	19
98262	54NU UMA	K3III	3.48		1.39	1.55	2.02	1.24	1	2.1				+1.9	SUN	-0.19	151
100006	86 LEO	[K0]	5.58		1.07	0.76	1.94	1.06	1	3.00				+1.71	SUN	+0.02	152
101013		[K0]	5.92		0.78		**	1.17	2		+2.27			+5.9	ALF BOO	+0.48	47
101013								1.19	3		-0.50				SUN	-0.33	126
101065		[B5]	8.01	*	0.74	0.23		0.80	3		+2.64			+3.3	SUN	+0.8	200
101501	61 UMA	G8V	5.35	5.55	1.18	1.16	**	0.92	1	4.60	+0.90	-0.89		+1.3	SUN	-0.14	61
101501								1.04	3					+5.0	SUN	-0.40	126
102224	63CHI UMA	K0III	3.72				1.78	1.26	2					+2.94	SUN	-0.65	49
102365		G5V	4.90	4.85	0.66	0.08		0.93	1	4.08		-0.89		+1.3	SUN	-0.48	86
102634		[F8]	6.15		0.52	0.11		0.83	1	4.3				+2.0	SUN	+0.14	50
102870	58ET VIR	F8V	3.61	3.60	0.55		0.90	0.84	4	4.29		-0.02	+0.05		SUN	+0.33	3
102870								0.82	1	4.12				+3	SUN	+0.26	59
102870								0.83	1					+1.3	SUN	+0.28	61
102870								0.83	4		+0.41	-0.04	+0.10		SUN	+0.15	75
102870								0.83	4						SUN	+0.19	85
103095	GMB 1830	68VP	6.45	6.71	0.75	0.17	0.96	1.12	2			-1.41			SUN	-1.50	35
103095								0.99	4			-0.58			SUN	-1.0	96

TABLE II (CONTINUED)

HD	NAME	SPECT.TYP	VM	MV	B-V	U-B	1-X	θ	1 LG G LG FE	[FE]	[Y]	ξ+	STANDARD	[FE/H]	REF
103578	95	LEU A3V	5.47		0.11	0.11		0.57 1 3.90					SUN	-0.01	11
103877		[AN]	7.2	*				0.70 4 4.0		+0.42 +0.41	+6.5	SUN	+0.66	33
103877								0.68 1 4.0				HK 114,906,2085,4825	+0.41	125
104304		[K0IV]	5.55	5.06	0.76	0.43		0.92 1 4.24					SUN	+0.18	86
104979	9UM1	VIR 88111	4.11		0.99	0.84	1.83	1.05 1 3.0		-0.12	+1.3	SUN	-0.6	186
105452	1ALF	CRV F2V	4.03	3.1	0.32	-0.02	-0.48	0.73 1 4.20			+2.5	EPS VIR	-0.60	191
105590		[3U]	6.54		0.84			0.89 1 3.90			+2.4	SUN	-0.04	90
106304		[8V]	9.07	*	0.03	-0.05		0.55 2	+2.40		+2.60	HD 186202	-0.8	201
											+1.9	ALF LYR		
106516		COM AM	6.11		0.45	-0.14		0.87 4 4.0		+0.24 -0.15		SUN	+0.05	3
106516								0.92 4 4.3		-0.72 -0.15		SUN	-0.86	3
106516								0.83 1				SUN	-0.40	50
107168	8	COM AM	6.27	*	0.17	0.14	-0.15	0.58 1 4.0			+7.2	HK 114,906,2085,4825	+0.83	123
107168								0.74 2	+1.12		+4.8	SUN	-0.09	230
107328	16	VIR K1111	4.95		1.16	1.15	0.73	1.32 2 1.72	-1.16		+1.98	6AM TAU	-0.72	43
107328								1.14 1 2.4		-0.03	+1.70	EPS VIR	-0.11	139
108381	156AM	COM K1111-IV	4.37		1.13	1.15	1.62	1.08 1 2.8		+0.07	+2.30	EPS VIR	-0.20	139
108486		AM	6.76	*	0.16	0.10		0.59 1 4.0			+6.0	HK 114,906,2085,4825	+0.39	123
108642		[AN]	6.54		0.18	0.11		0.60 1 4.0			+7.5	HK 114,906,2085,4825	+0.50	123
108651		AUF	5.85		0.22	0.08	**	0.60 1 4.0			+7.0	HK 114,906,2085,4825	+0.61	123
108662	17	COM AUF	5.29		-0.05	-0.12	**	0.43 1			+9.55	U PEG	-0.11	190
109307	22	COM A4V	5.29		0.11	0.09	**	0.56 1 4.0			+6.2	HK 114,906,2085,4825	+0.53	123
109358	8BE1	CVN G0V	4.29		0.59	0.05	1.16	0.88 2 4.35		-0.08 +0.03	+1.3	SUN	+0.02	19
109358								0.86 1		+0.04 +0.19		SUN	+0.08	62
109358								0.85 4				SUN	-0.23	75
109510	24	COM K2111+IV	6.72	4.46	0.25	0.11	**	0.64 1	+0.47			SUN	+0.54	154
109510	S							0.64 4				SUN	+0.42	154
109995		A0V	7.62	*	0.04	0.11		0.64 1 2.8	+1.37	+0.80 +0.54		ALF LYR	-1.12	36
109995								0.64 1 2.8	+1.37	+1.02 +0.25		ALF CMA	-1.26	36
109995												HD 161817	-0.3	69
109995												HD 161817	-0.12	72
110897	10	CVN G0V	5.96	4.7	0.55	-0.03		0.87 4 4.56		-0.15 -0.10		SUN	-0.32	3
110897								0.87 2			+0.3	SUN	-0.30	61
110897								0.89 4				SUN	-0.47	85
111153		A4P	6.35		-0.05	-0.08		0.52 1 3.25			+1.5	U PEG	+1.11	118

TABLE II (CONTINUED)

HD	NAME	SPECT.TYP	VK	MV	B-V	U-B	1-X	θ	1 Lg G Fe	Fe	[Fe]	[Y]	Tg+	STANDARD	[Fe/H]	REF
112033	35 COM	G8III+F6	4.87		0.90	0.66 **	1.59 *	1.19	2 2.71	-0.28			+1.88	GAM TAU	0.00	43
112127		K1III	6.92 *		1.27	1.42 **		1.10	1 2.3				+2.4	SUN	-0.09	151
112989	37 COM	K1P	4.90		1.17	1.05 **	1.92 *	1.21	2	-0.98			+5.08	SUN	-0.44	49
113226	47EPS VIR	G8III	2.81		0.94	0.74	0.95	1.02	1 2.7					SUN	+0.01	4
113226								1.02	1		-0.05			G.II+E TAU	-0.15	6
113226								1.14	1					SUN	+0.04	25
113226								1.02	1 2.7				+2.2	SUN	+0.04	41
113226								1.19	1 2.45	-0.46			+1.83	GAM TAU	-0.15	43
113226								1.02	1 2.70					SUN	+0.02	99
113226								1.01	1 2.85					SUN	+0.04	132
113226								1.01	1 3.00				+5.5	SUN	-0.1	138
113226								1.01	1 2.7				+1.8	SUN	0.00	220
113226								.021	1 2.0			+0.03		SUN	-0.06	224
114330	51THE VIR	A1V	4.37	4.66	-0.01	0.01	-0.05	0.53	1 4.0				+5	NORMAL A *	0.00	145
114710	43BET COM	G0V	4.28		0.58	0.08	0.99	0.83	4		+0.16			SUN	+0.19	3
114710								0.87	2		-0.01			SUN	+0.08	19
114710								0.85	1		+0.06			SUN	+0.05	37
114710								0.85	1 4.47				+1.3	SUN	+0.27	62
114710								0.84	4			+0.10		SUN	+0.16	74
114710								0.83	4					SUN	+0.18	85
114762		F9V	7.31 *	4.7	0.54	-0.07	**	0.86	4			-0.10	+5	SUN	-0.59	3
115043		G1V	6.83 *		0.60	0.08	**	0.85	4		-0.04	-0.05		SUN	-0.14	3
115043								0.97	3		+0.05	-0.05		SUN	-0.06	29
115383	59 VIR	G0VS	5.23	4.65	0.58	0.09		0.85	1 4.44				+1.3	SUN	+0.10	62
115604	20 CVN	F0II-111F	4.71		0.30	0.20	0.15	0.64	1 3.8				+3.5	SUN	+0.32	225
115604								0.67	1 4.1				+2.0	SUN	+0.44	226
116685	792ET UMA	AM	3.95		0.13	0.09	**	0.57	1 4.0				+7.2	HK 114,906,2085,4825	+0.24	123
116713		[KO]	5.25		1.21	1.02	**	0.98	2	-0.49			+5.3/	THE CEN ,ALF IND	+0.08	16
116713								1.08	1	-0.5/			+5.0	ALF BOO	+0.3	17
116713								1.03	1					ALF BOO	+0.3	177
117176	70 VIR	G5V	4.98	3.6	0.71	0.26		0.92	1 3.75				+1.3	SUN	-0.11	62
119796		[G81A-O]	6.41		1.84			1.03	1 0.00				+0.02	SUN	+0.02	131
120136	4TAU BOO	F7V	4.51	3.5	0.48	0.04		0.78	1 4.3				+2	SUN	+0.28	51
120709	3 CEN	B5III	4.72		-0.13	-0.60	0.18	0.26	1 3.87				+4	SOLAR SYSTEM	+0.67	161
120709								0.28	1 3.7				0	SOLAR SYSTEM	+0.53	162
120709								0.34	2	+2.27				GAM PEG	+0.63	192
121370	8ETA BOO	G0IV	2.69	2.72	0.58	0.20	0.58	0.85	1 3.8			-0.15	+0.78	EPS VIR	+0.53	139

TABLE II (CONTINUED)

HD	NAME	SPECT.TYP	VM	MV	B-V	U-B	1-X	θ	1	LG θ	G/LG FE	[FE]	[Y]	lg+	STANDARD	[FE/H]	REF
122563		G01V	6.20		0.90	0.38		1.18	4			-2.7	+0.00		SUN	-2.65	2
122563								1.24	1			-4.0	-0.05		SUN	-2.9	14
122563								1.09	1	1.2					SUN	-2.7	76
122563								1.10	1	1.2				+2.6	SUN	-2.72	79
122563								1.20	4			-3.0			SUN	-2.6	83
123139	5THE CEN	K0III	2.05	1.2	0.99	0.90	-1.29	1.10	2		-0.32		-0.10	+2.82	SUN	+0.30	16
123139								1.06	1	2.8				+1.23	EPS VIR	-0.19	139
123299	11ALP URA	A0III	3.84		-0.05	-0.08	0.10	0.55	1	2.5				0	SUN	-0.18	166
124425		F6IV	5.93		0.47	0.02		0.81	1		+1.25			+1.4	110 HER	+1.01	175
124425								0.79	1		+1.45			+1.4	110 HER	+1.62	175
124448		B3P	10.00	*	-0.09	-0.80	**	0.31	1	2.2	+2.43			10	B *	+0.2	196
124448								0.30	1	3.7				+7.0	SOLAR SYSTEM	+1.01	182
124897	16ALF B00	K2IIIP	0.06	-0.24	1.23	1.28	1.33	1.32	1	0.90					SUN	-0.30	21
124897								1.18	1	1.73					SUN	-0.30	34
124897								1.18	1					+1.8	SUN	-0.40	52
124897															SUN	-0.70	181
124897																-0.43	224
125162	19LAM B00	A0P	4.18	*	0.08	0.05	-0.05	0.60	1	4.0				+1.6	ALF LYR	+0.1	174
126661	22 B00	[A5]	5.27					0.65	1	3.5			0.00		SUN	+0.1	231
128167	28SIG B00	F2V	4.45	*	0.37	-0.08	0.41	0.75	1		+1.89			+2.0	ALF CMI	-0.6	170
128167								0.74	1	4.35				+2.9	SUN	-0.42	195
128279		[G0]	8.0	*	0.64			0.92	1	3.5				+0.6	SUN	-2.05	193
128620	A ALF CEN	G2V	0.33	4.35	0.60	0.76	**	0.87	4		-0.04	-0.01		SUN	+0.22	80
128621	B ALF CEN	K0V	1.70	5.69	0.85	0.70	1.65	0.94	4		-0.16	-0.03		SUN	+0.12	80
129174	29P1 1B00	B9P	4.54		-0.04	-0.32	-0.05	0.46	1				+/	6AM TAU	-0.29	160
129174								0.45	1	3.75			+3.02	U PEG	-0.64	190
129174								0.40	1	3.5			+1.9	COSMIC	-1.5	211
129312	31 B00	G8III	4.85		1.00	0.76	1.80	1.21	2.16	-0.71		+1.98	6AM TAU	-0.28	43
129312								1.05	1	2.1			+4	GAM,DEL,EPS TAU,EPS, VIR	0.00	138
130952	11 LIB	B8+III-	4.95		0.98	0.70	1.65	1.22	1	2.6/	-0.52	-0.05	+1.93	6AM TAU	-0.49	43
130952								1.03	1	2.9			+1.50	EPS VIR	+0.05	139
131156	37X1 B00	G8V	4.54	5.53	0.77	0.29	1.35	0.91	1	4.4	+0.02	+0.20	+2.3	SUN	0.00	19
131156								0.92	1						SUN	-0.26	151
132345	18 LIB	K3PIII-IV+B8	6.02		1.32	1.49	**	1.15	1	2.3			+2.3	SUN	-0.02	151
135485		B5P	8	*				0.27	1	4.3				0	SOLAR SYSTEM	+1.9	158
135485								0.27	1	4.3				+5	SOLAR SYSTEM	+1.4	158

TABLE II (CONTINUED)

HD	NAME	SPECT.TYP	VM	MV	B-V	U-B	|1-X	θ	1	L6	6	L6	HE [Fe]	[S]	ξ+	/ STANDARD	[Fe/H]	REF
135722	49BEL	BOU G0V+G8III	3.50		0.95	0.68	*	1.23	1	2.55		-0.86			+1.88	GAM TAU	-0.57	43
135722								1.01	1	2.4					+3.5	GAM,DEL,EPS TAU,EPS VIR	-0.4	138
136064		F8V	5.14	3.6	0.53	0.08		0.84	1	4.06					+1.3	SUN	-0.03	61
136202	5	SEK F8III-IV	5.06		0.54	0.06	**	0.83	1	3.89					+1.3	SUN	-0.06	62
136202								0.83	4			+0.38	-0.07			SUN	-0.17	75
136202								0.84	1	3.90					+2.8	SUN	0.00	90
136352	HU	2LUP G2V	5.62	4.70	0.65			0.92	1	3.92			-0.45		+1.1	SUN	-0.52	86
136352								0.92	1	4.18					+1.3	SUN	-0.46	208
136512	10MI	CRB K0III	5.51		1.02	0.77		1.23	2	2.64		-0.56			+1.91	GAM TAU	-0.32	43
137759	12IOT	DRA K2III	3.26		1.16	1.23	1.64	1.11	1	2.60					+1.4	EPS VIR	+0.30	140
137909	3BET	CRB F0P	3.66		0.29	0.11	0.15	0.65	2	4.5		+1.70			+4	2 HYA	+0.70	24
137909								0.64	2						+5.0		+1.01	228
138716	37	LIB K1III	4.62		1.01	0.85	1.59 *	1.06	1	3.2					+1.85	EPS VIR	-0.14	139
138905	386AM	LIB G8III-IV	3.90		1.02	0.74	1.54	1.08	1	2.9					+1.24	EPS VIR	-0.49	139
139195	16	SEK K0P	5.26		0.95	0.06	**	1.18	2	2.70		-0.29			+2.0	GAM TAU	-0.06	43
139195								1.23	3			-0.65			+4.2		-0.19	126
139669	15THE	UMI K5III	5.14		1.58	1.69	2.35 *	1.38	1	1.7					+1.9	SUN	+0.20	151
140232	22TAU	7SEK [A8V]	5.72 *					0.65	1	4.44					+3.6	SUN	+0.52	88
140283		F3VI	7.22		0.49	-0.20	**	0.85	1	4.6						SUN	-2.00	3
140283								0.80	1	4.80						SUN	-2.48	7
140283								0.94	2					-0.25		SUN	-1.04	11
140283								0.92	1	3.3					2-3	SUN	-2.03	38
																	-1.69	58
140573	24ALF	SER K2III	2.65	1.1	1.17	1.25		1.14	4				-0.10	+0.05		SUN	+0.08	34
140573								1.09	4				-0.06	+0.18	+1.8	SUN	+0.20	60
140573								1.11	1			+0.39	-0.07			SUN	+0.23	151
140573								1.26	1						+3	ALF BOO	+0.25	231
141004	27LAM	SER G0V	4.43	4.30	0.60	0.10	1.11	0.89	2							SUN	-0.04	19
141004								0.85	4					+0.06		SUN	+0.15	75
141004								0.85	1	3.94					+1.3	SUN	-0.02	86
141556	5CHI	LUP B9IV	3.94		-0.04	-0.14	0.09	0.45	1	3.1						SUN	-0.62	94
141714	10DEL	CRB G5III-IV	4.62		0.80	0.37	1.37 *	0.76	1	3.1			-0.10		+1.20	EPS VIR	-0.04	139
142198	46THE	LIB K0III-IV	4.14		1.01	0.82	1.60	1.06	1	2.9			+0.02		+1.96	EPS VIR	-0.24	139
142267	39	SEK G2V	6.10		0.60	0.00		0.87	4				-0.19	+0.02		SUN	-0.28	3
142373	1CHI	HER F9V	4.60	3.35	0.57	0.00		0.91	2				-0.41	+0.02		SUN	-0.40	19
142373								0.87	1			+0.36	-0.09	+0.09	+1.3	SUN	-0.29	61
142373								0.85	4	3.93						SUN	-0.35	75

TABLE II (CONTINUED)

HD	NAME	SPECT.TYP	VM	MV	B-V	U-B	I-X	θ	N	LG G	LG PE	[PE]	[Y]	ξ+	STANDARD	[FE/H]	REF
142860	41GAM SER	F6V	3.85	3.4	0.48	-0.03	0.95	0.79	1	4.0				0.9-6.5	SUN	-0.40	8
142860								0.82	1	4.0				+2	SUN	-0.07	55
142860								0.82	4						SUN	-0.11	75
143761	15RHO CRB	G2V	5.40	4.1	0.60	0.09		0.89	4						SUN	-0.20	3
143761								0.87	1	3.98		-0.25	-0.05	+1.3	SUN	-0.17	62
143761								0.86	4		+0.45	0.00	+0.18		SUN	-0.14	85
143807	14IOT CRB	A0P	5.02		-0.07	-0.20	0.08	0.46	1	3.7				+2.5	SOLAR SYSTEM	+0.17	93
143807								0.40	1	3.8					SUN	+0.8	94
143807								0.42	1	3.75				+3.02	U PEG	-0.18	190
144206	6UPS HER	B9P	4.75		-0.11	-0.32		0.50	1	3.5					SUN	+0.35	94
144941	[B8]		10.11 *		0.05	-0.71 **		0.23	1	3.5				00	SUN	-0.1	204
144941								0.23	1	3.5				10	SUN	-0.8	204
145389	11PHI HER	B9P	4.27		-0.07	-0.28 **	0.09	0.44	1	3.5		-0.25		+2.5	SUN	+0.30	94
145389								0.40	1	4.0					SUN	+0.2	146
145675	14 HER [K0V]		6.62 *		0.88	0.67 **		0.97	1	4.5				+1.4	SUN	+0.22	150
145675								0.97	1	4.4				+1.9	SUN	+0.18	151
146233	18 SCO [G1V]		5.49	4.42	0.65	0.17 **		0.86	1	4.18		-0.05		+1.3	SUN	+0.02	86
148816		F8V+	7.27 *		0.54	-0.07 **		0.91	4			-0.49	-0.07		SUN	-0.54	3
148856	27BET HER	G8III	2.83 *		0.93	0.67 **		1.20	2		+1.85				ALF BOO	+0.18	47
148856								1.14	2		+2.39				ALF BOO	+0.64	47
149438	23TAU SCO	B0V	2.82		-0.25	-1.01	0.37	0.15	1	4.1				+4.5	SUN	+0.8	105
149438								0.15	1	4.2				+5	SUN	+0.8	119
150680	40ZET HER	G0IV	2.82	2.97	0.65	0.21	0.68	0.87	2			-0.20	-0.03	+0.03	SUN	+0.07	37
150680								0.89	1	3.8			-0.15	+0.85	EPS VIR	-0.19	139
150997	44ETA HER	G8III	3.50	1.8	0.92	0.61	1.44	1.11	1				-0.10		SUN	0.00	25
150997								1.01	1	3.1				+1.25	EPS VIR	-0.21	139
151199		A5P	6.16 *		0.07	0.11	+1.46	0.58	1		+2.1			+1.30	EPS VIR	+0.01	82
151680	26EPS SCO	K2III	2.28	1.1	1.16	1.16		1.12	1	2.5			-0.01		SUN	-0.30	139
152792		G0V	6.82 *		0.64	0.08 **		0.90	4			-0.49	0.00		SUN	-0.45	3
152792								0.89	4						SUN	-0.31	85
153210	27KAP OPH	K2III	3.19		1.16	1.16	1.60	1.10	1	2.3				+1.5	SUN	+0.07	217
153210								1.10	1	2.3				+2.0	SUN	-0.06	217
153286		AM	6.88 *		0.33	0.20 **		0.67	1	3.5				+4.0	SUN	+0.17	137
154733		K4III	5.55 *		1.30	1.52 **	2.02								SUN	0.00	21
155646		F8IV	6.44 *		0.50	0.03 **		0.82	1	4.00				+3.50	HD 136202	+0.02	90

TABLE II (CONTINUED)

HD	NAME	SPECT.TYP	VM	MV	B-V	U-B	I-X	θ	1	LOG G	LOG FE	[FE]	(X)	ξ+	STANDARD	[Fe/H]	REF
155885	36	OPH K1V	5.33		0.86	0.49		0.99	1	4.6				-0.03		SUN	-0.01	221
155886	36	OPH K0V	5.29					0.99	1	4.6				-0.03		SUN	+0.09	221
156026		K5V	6.32 *	7.66	1.15	1.03 **		1.17	1	4.7					+1.3	SUN	0.00	150
156026								1.11	1	4.7					+2.0	SUN	-0.13	216
156074		K1(CL2)	7.60 *		1.14	0.92 **		1.06	1	2.05						SUN	+0.8	178
157089		F9V	6.95 *		0.59	0.00 **		0.89	4			-0.49				SUN	-0.57	3
157089								0.91	4							SUN	-0.54	85
157214	72	HER G0V	5.39		0.62	0.07		0.90	4	4.27		-0.34	-0.05			SUN	-0.36	3
157214								0.90	4						+1.3	SUN	-0.58	61
157214								0.90	4							SUN	-0.34	85
157999	49816	OPH K311	4.34	4.71	1.50	1.58	2.10 *	1.47	3		-1.20				+5.7	SUN	+0.01	126
158614		G8IV-V	5.34	4.58	0.72	0.31		0.90	1	4.39			-0.02		+1.3	SUN	+0.02	86
160691	MU ARA	G5V	5.12	4.9	0.70	-0.00		0.90	1	4.20					+1.3	SUN	+0.41	208
160693		G0V	8.37 *		0.60	0.00 **		0.91	4			-0.60	-0.10			SUN	-0.89	3
160762	85101	HER B3V	3.80		-0.18	-0.69 **		0.26	1	3.5						SUN	+0.71	94
160762								0.28	1	4.0						SUN	-0.4	104
160762								0.25	1	3.75						SUN	0.00	106
161227		F0II	8.5 *	3.89				0.67	1	3.5					+7.0	HR 114,906;2085,4825	-0.01	125
161797	86NU	HER G5IV	3.35		0.76	0.39 ***	1.14 *	0.91	1	3.91					+1.3	SUN	+0.16	61
161797								0.93	1	3.8				-0.15	+0.89	EPS VIR	+0.13	139
161817		A2VI	6.98 *		0.17	0.12 **		0.66	3		+1.				+5	SUN	-1.21	9
161817								0.55	4								-1.28	36
161817								0.67	2		+1.40					ALF CMA	-0.54	38
161817								0.66	1		+1.34				+2	SUN	-1.6	69
161817								0.64	1							ALF CMA	-0.98	141
162211	87	HER K2III	5.17		1.16	1.11 ***	1.86 *	1.10	1	2.80					+1.7	SUN	+0.07	152
164136	94NU	HER F2II	4.41 *		0.39	0.15	0.55	0.74	1	3.40					+5.4	SUN	-0.26	195
165195		[K3F]	7.35 *		1.1	0.68 **		1.22	1			-3.5	00.0			SUN	-2.7	14
165341	70	OPH K0V	4.02	5.67	0.86	0.51	1.45 *	1.08	3		+1.45				+2.8	SUN	-0.52	126
165341								0.95	1	4.5					+1.3	SUN	0.00	150
165341								0.95	1	4.4					+2.3	SUN	-0.12	151
165760	71	OPH G8III-IV	4.63		0.97	0.73	1.56 *	1.03	1	2.9				-0.02	+1.71	EPS VIR	-0.21	139
165908	99	HER F7V	5.04	4.02	0.52	-0.10		0.88	4	4.2		-0.44	-0.05			SUN	-0.51	3
165908								0.87	2	4.17		-0.40	0.00			SUN	-0.45	37
165908								0.84	1							SUN	-0.39	54
165908								0.85	1						+1.1	SUN	-0.42	62
165908								0.89	4						+1.3	SUN	-0.53	85

TABLE II (CONTINUED)

HD	NAME	SPECT.TYPE	VM	AV	B-V	U-B	1-X	θ	1 Lb G Lb PE	[PE]	[B]	ξ*	STANDARD	FE/H	REF
166208		G8III-(P)	4.99		0.92	0.69	1.38 **	1.15 1	2.83 -0.21			1.98	ETA HER IOT,KAP GEM,EPS VIR	+0.41	25
166208								1.16 1	2.83 -0.21			+4	GAM TAU	+0.02	43
166208								1.01 1	2.4				GAM,DEL EPS TAU,EPS VIR	+0.1	138
166620		K2V	6.40	6.24	0.87	0.59		1.04 1	4.5			+1.5	SUN	-0.20	150
166620								1.04 1	4.4			+2.3	SUN	-0.30	151
168322		G9III	6.10		1.00	0.70		1.17 1					IOT,KAP GEM,EPS VIR ETA HER	-0.21	25
168476		B5P	9.33 *		-0.01	-0.67 **		0.37 1	1.3	+1.58			B *	0.00	196
168476								0.38 2	2.4			14.5	B *	+0.85	182
168723	68E1A SER	K0III-1V	3.26	1.8	0.94	0.65	1.40 *	1.05 1	3.1		-0.17	+0.85	SOLAR SYSTEM EPS VIR	-0.31	139
168733	[B6]		5.39		-0.13	-0.58 **		0.31 1	3.7				SUN	+2.12	94
168733								0.35 1	3.6				SUN	+0.74	219
168785	[B]		8.5 *		-0.04	-0.74 **		0.21 1	3.3			+3.0	B *	-0.1	101
168913	HER A6V		5.50		0.21	0.04 **		0.64 1		+0.79	+0.25		SUN	+0.40	154
168913								0.66 1		+1.09	+0.23		SUN	+0.60	154
169027	[A0]		6.72 *		-0.07	-0.23 **		0.43 1	3.8			6	SUN	-0.31	67
169414	109	HER K2II	3.82		1.18	1.17	1.78 **	1.11 1	2.6		+0.05	+2.18	EPS VIR	-0.38	139
170153	44CHI	DRA F7V	3.58 *	4.13	0.49	-0.06	+0.75 *	0.82 1	4.3				SUN	-0.33	53
170153	5							1.02 1	4.5				SUN	-0.33	53
170820	[B6III]		7.37 *		1.57			1.07 1		-0.42	-0.20		6AM TAU	-0.08	12
170886	[G3III]		6.95 *		1.38	1.53		1.03 1		-0.95	-0.15		6AM TAU	-0.63	12
171443	ALF SCT	K3III	3.84		1.34	1.53	1.86 *						SUN	0.00	21
171653	[A6M]							0.67 1		+0.75	+0.44		SUN	+0.87	154
171653								0.67 1		+0.74	+0.37		SUN	+0.72	154
172167	3ALF LYR	A0V	0.04	0.50	0.00	0.00	0.08	0.53 1	4.0			+5.0	SUN	-0.05	65
172167								0.50 1	3.5			+2	SUN	-0.1	70
172167								0.56 1	3.8				SUN	+0.2	94
172167								0.55 1	4.0			+5.5	SUN	+0.06	97
172167								0.53 1	3.7			+5.0	SUN	-0.25	107
172167								0.53 1	3.7			+5.0	SUN	-0.1	117
172167								0.52 2	4.0	+0.5		+5.0	SUN	+0.02	124
172167													SUN	-0.1	129
172167													SUN	-0.3	155
173524	46	DRA [sy.5F]	5.04		-0.10	-0.29 **		0.42 1				+5.9	SUN	+0.4	197
173524	5							0.42 1				+5.9	SUN	+0.4	197
173648	6ZET 1LYR	AM	4.37		0.19	0.16		0.63 1	3.9	+1.12	+0.38	+6.0	SUN	+0.18	100
173648								0.60 1					SUN	+0.40	154

TABLE II (CONTINUED)

HD	NAME	SPECT.TYP	VM	AV	B-V	U-B	1-X	θ	1 LG G FE	[FE]	[U]	ξ+	STANDARD	[FE/H]	REF
173780		K3III	4.84		1.20	1.22	1.92						SUN	0.00	21
173880	11	HER A3V	4.36		0.12	0.08	-0.10	0.59				1-2	ALF LYR	-0.19	198
174638	10 BET	LYR B7V	3.43		0.00	-0.57	0.53	0.51 3	+1.10			+4	ALF CYG	+0.1	212
174704		[F-2]	7.72	*				0.68 1 3.5				+6.0	HR 114,906,2085,4825	+0.67	125
174704								0.84 2	+0.28			+4.2	SUN	+0.3	229
174933	12	HER B9II-III	5.20		-0.08	-0.42	**	0.37 1 4.0				+2.5	SUN	+0.62	148
174933								0.50 1 4.2				+6	SUN	+0.50	148
174933								0.36 1 3.75				+1.51	U PEG	+0.85	190
175329	UMG	PAV K1III-IV	5.13		1.37			1.25 1 2.0				+2.5	SUN	-0.58	208
175674		K3III	7.9	*				0.99 2	-0.42			+4.47	THE CEN+ALF INU	+0.25	16
176232	10	AQL A4F	5.90		0.25	0.08	**	0.62 1 4.0					SUN	+0.2	142
176232								0.71 2 4.18	+1.71				SUN	+0.01	144
178717		K5	7.47	*				1.00 2	-0.75			+6.31	THE CEN+ALF INU	+0.13	16
179761	21	AQL B8(P)	5.14		-0.07	-0.42	**	0.38 1 3.0				+3.02	U PEG	-0.41	190
180262		B8III-III	5.52		1.07	0.88	**	1.17 1					IOT KAP GEM EPS VIR, ETA HER	+0.21	25
180711	57 DEL	DRA G9III	3.07		1.00	0.78	1.06	1.02 1 3.00					SUN	+0.09	132
180928		[K2]	8.28		1.43	1.57	0.83	1.25 1 1.5					SUN	-0.60	222
181615	A6DPS	SGR A6FE	4.61	4.0	0.10	-0.53	2.00	0.77 3	-0.1		-0.02	+6	ALF CYG	0.00	212
182490	2	SGE A2III(*)	5.96		0.07	0.04	*	0.54 1	+0.84	-0.02		SUN	+0.64	154
182490	S							0.54 1	+1.05	+0.26		SUN	+0.97	154
1825/2	31	AQL G8IV	5.17		0.78	0.42		0.89 4 4.26	+0.08		+1.5	SUN	+0.50	86
1825/2								0.89 4	+1.59			SUN	+0.51	87
1825/2								0.92 1 4.0			+0.93	EPS VIR	+0.42	120
1825/2								0.87 1 4.13	-0.14		+1.9	SUN	+0.40	139
1825/2								0.88 1				SUN	+0.44	208
183915		[K+]	9.0	*				0.96 2	-0.30		+3.55	THE CEN+ALF INU	+0.30	16
184406	58 MU	AQL K3III	4.44		1.18	1.24	1.80	1.30 1 2.17	-0.61		+1.91	GAM TAU	+0.24	43
184406								1.07 1 2.7			+4	GAM,DEL,EPS TAU,EPS VIR	+0.3	138
185144	61 SIG	DRA N0V	4.68		0.80	0.57	0.85	0.97 1 4.60			+1.3	SUN	-0.23	62
185144								0.98 1 4.4			+2.0	SUN	-0.25	151
185395	13 THE	CYG F4V	4.48	3.2	0.39	-0.02	-0.56	0.78 4	+0.11	+0.1		SUN	+0.04	64
185657		[KU]	6.35					1.09 2	-1.47			SUN	-0.51	5
186408	16	CYG G2V	5.96	4.3	0.64	0.20	0.65	0.87 4	-0.03	+0.05		SUN	+0.22	3
186408								0.87 4				SUN	+0.20	85

TABLE II (CONTINUED)

HD	NAME	SPECT.TYP	VM	MV	B-V	U-B	I-X	θ	l	LG G [LG FE]	[Fe]	[V]	ξt	STANDARD	[Fe/H]	REF
186427		G5V	6.20	4.6	0.66	0.21	0.77	0.86	4		-0.04	+0.07		SUN	+0.11	3
186427								0.87	4					SUN	+0.07	85
186791	50GAM AQL	K3II	2.71		1.52	1.68	1.49							SUN	0.00	21
187923		G2V	5.78	*	0.53	0.21		0.89	4		-0.29	+0.07		SUN	0.00	3
187923								0.87	2		-0.16	+0.07		SUN	+0.12	31
188056	20	CYG K3III	5.02		1.29	1.51	1.99	1.34	2	1.79 -1.04			+1.99	6AH TAU	-0.14	43
188056								1.12	1	1.9			+4.5	GAM,DEL,EPS TAU,	-0.1	138
188512	60BET AQL	G8IV	3.71		0.86	0.49	1.34	0.98	1	3.6		+0.17	+0.82	EPS VIR	-0.18	139
188512								0.98	1		+0.43		+0.8	EPS VIR	-0.23	184
186650		[G5]	5.63		0.75			0.85	1	2.9				HD 204867	-0.4	232
188947	21E1A	CYG K0III	3.93		1.03	0.88	1.50	0.88	1	4.08			+0.9	SUN	0.00	21
189567		G2V	6.06		0.64	0.08		0.88	1	4.08		-0.22		SUN	-0.28	86
189849	15	VUL AM	4.65		0.18	0.16	-0.25	0.63	1	3.5				SUN	-0.02	73
189849								0.74	2	+1.12			+3.8	SUN	+0.02	230
190229		K8II-III	5.50	4.76	-0.11	-0.49		0.57	1	3.7				SUN	+1.05	94
190248	DEL PAV	G5IV-V	3.55		0.76	0.44		1.00	2	+0.57		+0.06		SUN	-0.02	113
190248								0.88	4					SUN	+0.29	135
190248								0.90	1	4.61			+1.3	SUN	+0.43	208
190360		G6IV+IM6	5.70	4.3	0.72	0.38		0.90	1	4.07			+1.3	SUN	+0.26	61
190404		K1V	7.28	5.9	0.82	0.39		1.10	3				+3.2	SUN	-0.20	39
190404								0.97	1	4.50			+1.9	SUN	-0.10	41
190404								1.03	1	4.5			+1.25	SUN	-0.15	150
190404								1.03	1	4.5			+1.1	SUN	-0.3	216
190404								0.97	1	4.50		+0.04		SUN	-0.14	224
191046		G9III	7.03	*	1.15	1.03		1.17	1					L,K GEM,E VIR,H HER	-0.42	25
191408		K3V	5.32		0.87	0.64	0.49	1.03	1	4.6		-0.03		SUN	-0.07	221
192310		K0V	5.73	6.13	0.88	0.64	-1.16	1.01	1	4.5			+1.3	SUN	-0.04	150
192310								1.01	1	4.5			+1.6	SUN	-0.11	216
192640	29	CYG A2V	4.99		0.12	-0.01	-0.27	0.63	1	3.9 +1.3			+4.0	ALF LYR	-0.5	174
192947	6ALF 2CAP	G9III	3.55		0.95	0.69	1.37	1.01	1	3.30			+0.6	SUN	+0.14	152
193370	35	CYG F5IB	5.22		0.65	0.47	0.42	0.89	2	-0.45			+5.5	SUN	0.00	229
193432	8NU	CAP B9V	4.75	*	-0.04	-0.11	0.05	0.49	1	3.75			+2.75	SUN	-0.12	110
193432								0.48	1	3.75			+2.75	SUN	-0.1	142
193664		G5V	5.94	5.07	0.58	0.06		0.84	1	4.64			+0.7	SUN	+0.06	62

TABLE II (CONTINUED)

HD	NAME	SPECT.TYP	MV	VM	B-V	U-B	1-X	Ð	1	LG G	LG FE	[FE]	[V]	ξ+	STANDARD	[FE/H]	REF
194093	376AM CYG	F8IB		2.23	0.67	0.54	0.68	0.92	1		+0.06			+6.3	SUN	+0.4	167
194093								0.94	2		-1.05				SUN	+0.1	229
195295	41 CYG	F5II		4.02	0.40	0.30	0.85 *	0.75	1	2.50				+7.1	SUN	-0.05	195
195725	21HE CEP	AM		4.21	0.20	0.16	-0.26	0.60	1	4.0				+7.8	HK 114+906+2085+482...	+0.13	123
195725								0.72	2		+1.56			+7.6	NORMAL *	+0.14	143
196171	ALF IND	K0III	1.1	3.10	1.00	0.79		1.12	2					+3.98	SUN	+0.25	16
196662	14TAU CAP	B6III		5.30 *				0.26	1	3.5					SUN	+0.80	94
196755	7KAP DEL	G5IV+UK1		5.02 *	0.74	0.19		0.89	1	3.7			-0.11	+1.07	EP'S VIR	+0.45	139
196777	15UP'S CAP	M2III		5.10 *	1.64	1.99	2.58	0.46	1	4.0					SUN	+0.1	197
197345	50ALF CYG	A2IA		1.26 *	0.09	-0.23		0.63	1		+0.05			+9.7	SUN	-0.04	114
197345								0.55	1	1.15	+0.48				SUN	+1.00	189
197461	11 DEL DEL	A7III	0.8	4.53	0.32	0.10	0.17	0.71	1	3.8				+7	SUN	-0.19	227
197989	53EP'S CYG	K0III		2.45 *	1.03	0.87	1.08	1.16	4	2.85					SUN	-0.25	34
197989								1.03	1						SUN	+0.08	132
198149	3ETA CEP	K0IV	2.72	3.43	0.92	0.61	1.40	1.01	1	2.95					SUN	-0.12	132
198149								1.01	1	3.4			-0.20	+0.66	EP'S VIR	-0.21	139
198269		K0		8.23 *	1.34			1.13	1			-0.70	-0.21		SUN	-1.56	108
200465		K3II-III+A1V		7.6	1.53	1.80	1.65	1.48	3		-1.20			+4.8	SUN	-0.16	126
200905	62X1 CYG	K5IB		3.72	1.65	1.11	0.65	1.29	1	0.75				+2.8	SUN	+0.06	132
201091	61 CYG	K5V	8.39	5.19	1.17	1.23	1.30	1.15	1	4.7				+2.0	SUN	0.00	216
201092	61 CYG	K7V		6.02	1.37	0.09	-0.36	1.38	1	4.6				+2.2	SUN	-0.65	151
201601	56AM EQU	F0P		4.66	0.26		1.41	0.67	1	3.5	+1.65		-0.20	+7.59	U PEG	-0.01	190
201601								0.74	2					+0.74	SUN	+0.39	241
201626		[G9P]		8.14 *	1.11	0.49		1.02	4			-0.7			SUN	-1.45	18
202109	64ZET CYG	G8II		3.20	0.99	0.76	0.77	0.98	4			-0.23		+2.7	EP'S VIR	-0.1	102
203608	GAM PAV	F6V	4.53	4.22 *	0.48	-0.13	-0.77	0.87	1	4.0		-0.54		+1.42	SUN	-0.67	45
203608								0.89	4	4.3					SUN	-0.7	71
204075	34ZET CAP	[GF]		3.73	1.00	0.60	1.17	0.85	2		-0.51			+3.98	THE CEN+ALF IND	+0.18	16
204411		AP		5.27	0.07	0.16		0.59		4.0				+4.0	SUN	-0.4	130
204411								0.58		4.3				+3.5	SUN	-0.6	130
204411								0.55	1	3.0				+3.37	U PEG	-0.07	190
204867	22 BET AQR	G0IB		2.89	0.84	0.58	0.41	0.93	1	1.4					SUN	0.00	232

TABLE II (CONTINUED)

HD	NAME	SPECT.TYP	VM	MV	B-V	U-B	1-X	θ	i	LG G	LG FE	[FE]	[X]	ξ+	STANDARD	[FE,H]	REF
205512	72 CYG	K1III	4.91		1.08	1.00	1.62 *	1.27	2	2.27	-0.65			+1.98	GAM TAU	-0.14	43
206088	406AM CAP	F0P	3.66		0.32	0.21	0.63 *	0.75	1	4.0	2.15			+5.7	30 LMI	+0.60	194
206088								0.67	1	2.56				4.9-6.7	SUN	-1.23	210
205546	P	AM	6.22		0.27			0.66	3				+0.75 +0.41		SUN	+0.78	154
205546	S							0.60	1				+1.19 +0.28		SUN	+0.90	154
206778	8EPS PEG	K2IB	2.42		1.52	1.70	1.39	1.21	1	1.0				+5.5	ALF BOO	+0.37	68
206778								1.19	1	1.25						-0.03	112
206778								1.32	3		-2.55			+7.6	SUN	-0.06	126
206778								1.19	1	1.10				+5.5	SUN	-0.05	132
206778								1.22	1						ALF BOO	+0.4	177
206859	9 PEG	G5IB	4.35		1.18	0.96	1.65 *	1.04	1	1.75				+5.4	SUN	+0.02	132
206952	11 CEP	K0III	4.57		1.10	1.09	1.69 *	1.08	1	2.5				+1.5	SUN	+0.05	151
207673		A2IB	6.42 *		-0.07	-0.44 **		0.57	1	1.04	+1.25			+/	ALF CYG	-0.16	172
207857		B8IIIP*	6.16 *		0.42	-0.12 **		0.38	1	3.8					SUN	+0.67	94
207978	15 PEG	F5V	5.51 *		0.60	0.08 **		0.82	1	3.8			+0.03		SUN	-0.69	215
208776		G0V	6.95 *		0.50	-0.11 **		0.85	1	4.0				+5.30	HD 136202	-0.14	90
208906		F8V	6.96 *		0.23	0.15	-0.28	0.86	1	4.0				+2.20	HD 136202	-0.51	90
209625	32 AQR	AM	5.32		0.97	0.77	0.84	0.61	1	4.0				+9.0	HR 114,906,2085,4825	+0.57	123
209750	34ALF AQR	G2IB	2.93		0.34	0.09	-0.45	0.99	1	1.45				+5.9	SUN	+0.03	132
209791	17X1 CEP	AM	4.29		0.44	-0.03	0.99 *	0.67	1	4.0				+7.0	HR 114,906,2085,4825	+0.32	125
210027	24IOT PEG	F5V	3.76		0.43	0.24		0.88	2			-0.34	-0.04		SUN	-0.10	19
210221		A3IB	6.14		1.55	1.72	1.38	0.53	1	1.50	+1.29			+/	ALF CYG	+0.17	172
210745	21ZET CEP	K1IB	3.36		0.99	0.80	1.43	1.32	3		-2.55			+8.2	SUN	-0.10	126
211391	43THE AQR	G8III-IV	4.16					1.02	1	2.8			-0.05	+1.53	EPS VIR	-0.07	139
211594	[K0F]		9.3 *	4.3	0.65	-0.06		0.95	2		-0.58			+3.89	THE CEN+ALF IND	-0.02	16
211998	NU IND	G0V	5.28					1.03	4	4.0					SUN	-1.17	56
211998								1.01	4						SUN	-1.3	71
211998								0.97	1	3.77				+1.3	SUN	-1.54	208
212061	486AM AQR	A0V	3.84		-0.06	-0.13		0.48	1	4.0	+2.4			+5.0	ALF LYR	+0.3	174
213009	BEL 16RU	G5	3.96		1.03	0.81	+1.34	1.15	1	0.90				+1.3	SUN	-0.48	152
214539		B9V	7.22 *		0.02		**	0.65	2		+0.25			+5.0	ETA LEG+13 MON	-1.2	115

TABLE II (CONTINUED)

HD	NAME	SPECT.TYP	VM	MV	B-V	U-B	1-X	θ 1 |L6 6 |L6 PL	[Pt]	[y]	Z+	STANDARD	[Fe/H]	REF
214714		GF	6.02		0.01	-0.01	-0.02	0.93 1 2.9				HD 204867	-0.4	232
214994	430M1	PEG A11V	4.79		0.01			0.50 1 4.00			+5.00	SUN	+0.04	127
214994								0.53 1 4.0			+5	SUN	+0.1	142
214994								0.45 1 4.0				NORMAL A *	+0.2	145
214994											+2.40	SUN	0.00	190
215104	KHO	GRU [KU]	4.84	2.6	1.03	0.81	1.41	1.20 1 1.00			+1.00	SUN	-0.50	152
215648	46X1	PEG F7V	4.19		0.50	-0.02	0.99	0.89 2		+0.02		SUN	-0.05	19
216228	32101	CEP K0111	3.53		1.05	0.90	1.53	1.01 1 3.55	-0.34			SUN	+0.09	132
216385	49S16	PEG F7V	5.16		0.48	-0.01		0.83 1 3.97			+0.5	SUN	-0.62	61
216735	56RHO	PEG A1V	4.89		0.00	0.00	-0.02	0.53 1 3.5 +2.0			+4.0	ALF LYR	+0.1	174
216763	23DEL	PSA K0	4.20	4.82	0.97	0.70	1.38	1.16 1 1.10	-0.08		+0.7	SUN	-0.56	152
217014	51	PEG G4V	5.53		0.67	0.20	0.63	0.88 1 4.27			+1.1	SUN	+0.12	86
217906	53BET	PEG M2II-111	2.56		1.67	1.96		1.7 3	-3.0			SUN	-0.85	168
218470	5	AND 68111	5.73		0.42	-0.01 **	1.41	0.81 1 3.80			+2.5	SUN	-0.31	91
218658	33P1	CEP G2111	4.42	6.41	0.82	0.45		1.18 2 2.67 -0.36			+1.86	6AM TAU	-0.30	43
219134		K3V	5.57 *		1.00	0.89	0.94	1.15 3 4.50			+4.4	SUN	0.00	39
219134								1.07 1 4.5			+2.9	SUN	+0.10	41
219134								1.08 1 4.5			+1.5	SUN	0.00	150
219134								1.08 1 4.4			+2.2	SUN	-0.21	151
219134								1.07 1 4.50			+2.0	SUN	0.00	216
219134								1.07		+0.18		SUN	-0.01	224
219615	68AM	PSC 67111	3.69		0.91	0.57	1.49	1.12 1			+7.59	IOT,KAP GEM,EPS VIR ETA HER UMI PEG	-0.16	25
219615								0.49 1 3.75					-0.18	190
219617		F81V	8.16 *		0.47	-0.20 **				-0.25		SUN	-1.40	3
219617													-1.40	38
219623		[+8]	5.59 *		0.54	0.02 **		0.89 2	+0.91		+1.0	SUN	-0.10	61
221148		K3111	6.25		1.08	1.14	1.06	0.83 1 4.07			+1.5	SUN	+0.07	151
221170		G2V	7.67 *		1.09	0.62 **		1.07 1 2.6	-3.5			SUN	-2.7	14
221345	14	AND K0111	5.22		1.02	0.86	1.78	1.23 1		-0.10		SUN	-0.20	21
222107	16LAM	AND 68111-1V	3.88		1.02	0.70		1.25 2 2.30 -0.79			+1.88	6AM TAU	-0.78	43
222368	17101	PSC F7V	4.13	3.39	0.51	0.00	1.06	0.84 1 3.96 +0.63	+0.18	+0.09	+0.8	SUN	-0.51	61
222368								0.82 4				SUN	+0.09	75
222404	356AM	CEP K11V	3.22	2.27	1.03	0.95	1.26	1.14 3 3.3 +1.20			+4.9	SUN	-0.04	126
222404								1.04 1 3.3		-0.06	+1.49	EPS VIR	-0.21	139

TABLE II (CONTINUED)

HD	NAME	SPECT.TYP	VM	MV	B-V	U-B	I-X	θ	1[LG G|LG PE]	[FE]	[V]	ξ+	STANDARD
223385	6	CAS A31A+	5.42		0.66	-0.02		0.54	1 1.00 +0.89			3.3 -10.6	ALF CYG
224635	85	[8]	6.58		0.52	-0.04		0.83	1 4.92			+0.7	SUN
224930		PEG G2V	5.75	5.38	0.67	0.05		0.95	4	-0.55	+0.05		SUN
224930								1.08	3	-0.63	+0.03		SUN
224930								0.94	2	-0.55	+0.10		SUN
224930							**	0.95	4	-0.58	+0.05		SUN
224930								0.97	1 4.35			+0.7	SUN
224930								0.92	4				SUN
225212	3	CET K31B	5.16		1.66		**	1.23	1 1.10			+3.0	SUN
232078		K311P			2.04	2.15	**	1.25	2	-3.09		+4.3	SUN
232078								1.19	4	-3.2	+0.34		SUN
NGC 752 205					+0.42	+0.01		0.75	1		+0.14	+2.6	SUN
NGC 752 213					+0.42	+0.01		0.76	1 3.6				SUN
NGC 752 218					+0.46	+0.04		1.04	1	-0.44	+0.07	+3.1	GAM TAU
NGC 2281 18		K0III						0.75	1				SUN
NGC 2281 63		K2III			+1.14	+1.09		1.12	1	-0.63	+0.05		GAM TAU
NGC 2548 8					+1.34	+1.41		1.17	1	-0.92	+0.07		GAM TAU
NGC 6633 156								1.12	1	-0.91	+0.10		GAM TAU
BD -48.8597								1.07	1	-0.70	0.00		GAM TAU
BD +10.2179			10					1.29	2 1.86 -1.14			+2.03	GAM TAU
BD +17.734		K0V						0.32	1 2.8			0	SUN
BD +39.708		K2V						0.98	1 4.44		+0.06		HD 344
BD +23.00551		G0V	9					1.04	1 4.44		+0.06		HD 344
BD +59.4926		A-+	8.4		-0.11			0.84	1			+0.25	SUN
CPH -69.2698		K1II			+1.68	+1.90		0.72	1 0.5			+5.1	SUN
								0.67	1 1.20			+0.8	SUN
K 13 140		K1II	12.03		+1.52	+1.86		0.17	1 4.1			+1.9	SUN
K 41 21								1.25	2	-2.99		+6.0	SUN
K 67 IV-202		K1III	11.2		+1.33	+1.11		1.15	2	-1.76		+3	ALP BOO
K 92 III-13								1.26	1			+1.2	SUN
								1.4	2	-4.79	+0.1		SUN
								1.25	4	-3.4			SUN
WILSON 10367		G5			+0.71	-0.12		0.92	1 4.44			+1.3	SUN

TABLE III

REF	DISPERSION	INTERVAL	REF	DISPERSION	INTERVAL
1	9.1-17.8	4000-4900	73	4.8-18.1	3700-6800
3	15	5160-6270	74	8	5180-5880
4	1 - 2.8	4500-6700	75	6.8-10	3600-6800
5	4.5- 9		76	9	3250-4800
6	6.0- 6.7	5100-6700	77	8	
7	10 -20	3430-8750	78	2.2	4760-5150
8	4.6-20	3800-8700	79	4.5- 8	3400-6700
9	10 -15	3300-6700	80	2.5- 6.7	4200-6900
10	20	3640-4420	81	10.2	3700-4900
11	10 -20	4000-4800	82	18	
12	13.5-16	5000-6300	83	6.7	4100-4900
13	6.5	5700-6500	84	1 - 1.5	4000-7500
14	4.5-15.3		85	15	
15	10 -15		86	6.7-10.2	4460-6800
16	6.8-15.6		87	6.7-10.2	4468-6810
17	6.7-20.4	4300-8700	88	9.9	
18	6.7-13.5	4300-6800	89	9 -18	
19	4 -16		90	10 -15	3900-6700
20	2.7		91	7.2-12.4	
21	10.2	4100-4500	92	4.5	
22	4.5		93	2 -8	3400-6700
23	9 -10	3800-4800	94	2.2- 8	3200-6600
24	2.1- 4.2		95	2	3850-4700
25	2.8-10	4170-4440	96	2.8- 6.8	3800-6600
27	10	4350-4850	97	2 - 3.7	
28	15	5167-6277	98	2 - 7	
29	13.5-15		99	2.0-12.4	
30	7 -14		100	9.7-39	3900-6300
31	15		101	12.3	3700-4800
32	11 -15		102	4.5- 6.7	
33	8		103	10.2	3700-4924
34	1.7- 4.2	4000-5000	104	2 -8	3400-6650
36	4.4-16	3800-6600	105	1.0- 6.5	
37	2.8-20.4	4000-8700	106	1.0-15.6	3448-6678
38	4.5-10		107	0.8-10	3100-8863
39	9		108	6.5-17.8	
40	15	5167-6277	109	0.8-10	3100-8868
41	9.7-12.4		110	4.5-10.4	
42	9 -14		111	4.6-15	
43	6.6	5200-6300	112	6.5	4400-6650
44	4.5	4650-6250	113	2.5	5600-6800
45	6.7	4000-4900	114	10.2	3700-4900
46	10		115	10.2-20.4	
47	6.7	5200-6300	116	2.8-10	3150-6565
50	4.5-15	3900-4650	117	0.8-10	3100-8863
51	3.2-12	4340-6700	118	12.4	3600-5000
52	1 - 1.6	5000-7025	119	1	3995-4920
53	3.2-4	4340-6750	120	2.9-12.3	
54	3 -12		121	2.9-12.9	
55	10		122	2.2-17.5	3000-6830
56	6.7-13		123	8.9-17.8	
57	6.8	4680-6770	124	2.7-16	
58	10		125	8.0-17.8	
59	4.3-15	4200-6800	126	12	3900-4600
60	1.6	5640-6320	128	8	5200-6240
61	7		129	10.2	3650-5000
62	7		130	4.5	3700-4900
63	6.7-10.2	4300-4800	131	13.4	
64	2 -8	3100-3160	132	1.6- 6.5	5000-6650
65	4.8-18.1	3700-6800	133	1.3	3700-4860
66	8 -20		134	8	3700-4600
67	4.5-9		135	6.7-10	4200-6758
68	1.5	5600-6300	136	10	4118-4630
69	9 -15	3300-6700	137	12.4	
70	10		138	8	5200-6200
72	9		139	20	5200-6400

TABLE III (CONTINUED)

REF	DISPERSION	INTERVAL	REF	DISPERSION	INTERVAL
140	2.2	4720-5240	185	12.3	3390-4930
141	10 -15	3800-6500	186	20	5950-6650
142	4.5	4200-4635	187	3.6-11	
143	8		188	9.4	
144	2.3- 6.9	3300-6600	189	2.8-20	3400-8860
145	4.5- 6.7		190	10	3200-3520
146	2 -8	3300-6400	191	3.2-12.4	4390-6700
147	9		192	4.5-13.5	3600-8600
148	2		193	12	
149	9.7		194	9.7-12.3	3600-6600
150	6.4-16.3		196	20	3700-4800
151	6.4-16.3		197	4	
152	3.0-12.4	4400-6300	198	15	3700-4750
153	2.7		199	16	3230-4645
154	8 -10		200	10.2	5257-6764
155	20.4	3350-5900	201	10.2	
156	2.8	4000-4800	202	2 -8	3100-6460
157	13.5-27		203	12.3	3700-4750
158	15.5	3600-4800	204	12.4-20	3100-5000
159	0.8-10	3100-8863	206	7 -20	3170-7800
160	4.5-15		207	12.3-31.3	3500-6600
161	4.5-13.5	3120-8680	208	6.7-10.2	
162	2	3732-4659	209	2.2	3400-6600
163	3 -8	3921-4623	210	2.8- 5.7	3736-6000
164	13	5350-6300	211	0.3-20.5	3734-4800
166	9.5	3700-4400	213	10.2	3700-4950
168	4.0	4090-4515	214	2	4425-4580
169	4.4-15	3200-4900	215	7.4-12.4	4000-6250
170	4.8-18.1	3700-6800	216	10 -15	3600-6800
171	9		218	9 -20	
172	9.7-12.4		219	2 -8	3456-6000
173	1.6	4500-6500	221	12.4	
174	4.5-20		223	10.5-16	
175	13.5		225	3.9-20.7	3300-8800
176	3	3900-4500	226	6.8-10	4200-4950
177	1.5	5000-7000	227	10 -15	3300-7800
178	6.7	5000-5900	229	10	
179	2	3677-4756	230	2.8- 8.6	3799-6588
180	5.6	4070-4510	231	9.7-12.4	3900-6150
181	0.9- 1.7	3600-5000	232	4.4-15.3	3800-6800
182	15.6-29	3100-6700	233	4.5	3970-4660
183	2	4425-4580	234	9	3300-4950
184	9 -18	3497-4144	235	10 -15	3900-6800

REFERENCES TO THE CATALOGUE

1 HELFER,H.L.,WALLERSTEIN,G. AND GREENSTEIN,J.L. 1959,ASTROPHYS.J. 129,700.
2 PAGEL,B.E.J. 1965,ROY.OBSERV.BULL. 104.
3 WALLERSTEIN,G. 1961,ASTROPHYS.J.SUPPL. 6,407.
4 CAYREL,G. AND CAYREL,R. 1963,ASTROPHYS.J. 137,431.
5 PAGEL,B.E.J. 1964,ROY.OBSERV.BULL. 87.
6 HELFER,H.L. AND WALLERSTEIN,G. 1964,ASTROPHYS.J.SUPPL. 9,81.
7 BASCHEK,B. 1959,Z.ASTROPHYS. 48,95.
8 KEGEL,W.H. 1962,Z.ASTROPHYS. 55,221.
9 KODAIRA,K. 1964,Z.ASTROPHYS. 59,138.
10 BASCHEK,B. 1965,Z.ASTROPHYS. 61,27.
11 CHAMBERLAIN,J. AND ALLER,L.H. 1951,ASTROPHYS.J. 114,52.
12 WALLERSTEIN,G. AND CONTI,P.S. 1964,ASTROPHYS.J. 140,858.
13 HAZELHURST,J. 1963,OBSERVATORY 83,128.
14 WALLERSTEIN,G.,GREENSTEIN,J.L.,PARKER,R.,HELFER,H.L. AND ALLER,L.H. 1963,
   ASTROPHYS.J. 137,280.
15 BURBIDGE,E.M. AND BURBIDGE,G.R. 1957,ASTROPHYS.J. 126,357.
16 WARNER,B. 1965,MONTHLY NOTICES ROY.ASTRON.SOC. 129,263.
17 DANZIGER,I.J. 1965,MONTHLY NOTICES ROY.ASTRON.SOC. 131,51.
18 WALLERSTEIN,G. AND GREENSTEIN,J.L. 1964,ASTROPHYS.J. 139,1163.
19 HERBIG,G.H. 1965,ASTROPHYS.J. 141,588.
20 RODGERS,A.W. AND BELL,R.A. 1963,OBSERVATORY 83,79.
21 SCHWARZSCHILD,M.,SCHWARZSCHILD,B.,SEARLE,L. AND MELTZER,A. 1957,ASTROPHYS.
   J. 125,123.
22 PAGEL,B.E.J. 1963,OBSERVATORY 83,133.
23 CHAFFEE,F.H.,CARBON,D.F. AND STROM,S.E. 1971,ASTROPHYS.J. 166,593.
24 HACK,M. 1958,MEM.SOC.ITAL. 29,263.
25 GREENSTEIN,J.L. AND KEENAN,P.C. 1958,ASTROPHYS.J. 127,172.
26 PAGEL,B.E.J. 1963,J.QUANTIT.SPECTROSC.RADIAT. 3,139.
27 HEISER,A.M. 1960,ASTROPHYS.J. 132,506. (CONTR.MCDONALD OBS.NO.327)
28 WALLERSTEIN,G. AND HELFER,H.L. 1959,ASTROPHYS.J. 129,347.
29 HELFER,H.L.,WALLERSTEIN,G. AND GREENSTEIN,J.L. 1960,ASTROPHYS.J. 132,553.
30 PARKER,R.,GREENSTEIN,J.L.,HELFER,H.L. AND WALLERSTEIN,G. 1961,ASTROPHYS.J.
   133,101.
31 WALLERSTEIN,G. AND HELFER,H.L. 1961,ASTROPHYS.J. 133,562.
32 KONDO,M. 1957,PUBL.ASTRON.SOC.JAPAN 9,201.
33 STICKLAND,D.J. 1972,MONTHLY NOTICES ROY.ASTRON.SOC. 159,29P.
34 GRATTON,L. 1953,MEM.SOC.ROY.SCI.LIEGE 14,419.
35 JUGAKU,J. (PRIVATE COMMUNICATION)
36 WALLERSTEIN,G. AND HUNZIKER,W. 1964,ASTROPHYS.J. 140,214.
37 HELFER,H.L.,WALLERSTEIN,G. AND GREENSTEIN,J.L. 1963,ASTROPHYS.J. 138,97.
38 ALLER,L.H. AND GREENSTEIN,J.L. 1960,ASTROPHYS.J.SUPPL. 5,139.
39 CAYREL DE STROBEL,G. 1964 IAU SYMP. 26 294 .
40 WALLERSTEIN,G. AND HELFER,H.L. 1959,ASTROPHYS.J. 129,720.
41 CAYREL DE STROBEL,G. 1966,ANN.ASTRON. 29,413.
42 KOELBLOED,D. 1967,ASTROPHYS.J. 149,299.
43 HELFER,H.L. AND WALLERSTEIN,G. 1968,ASTROPHYS.J.SUPPL. 16,1.
44 COHEN,J.G. 1968,ASTROPHYS.J. 154,179.
45 DANZIGER,I.J. 1966,ASTROPHYS.J. 143,527.
46 PAGEL,B.E.J. AND POWELL,A.L.T. 1966,ROY.OBS.BULL.124.
47 GRIFFIN,R. 1969,MONTHLY NOTICES ROY.ASTRON.SOC. 143,381.
48 EDMONDS,F.N.JR. 1965,ASTROPHYS.J. 142,278.
49 YAMASHITA,Y. 1964,PUBL.DOM.ASTROPHYS.OBS.VICTORIA 12,455.
50 CAYREL DE STROBEL,G. 1968,ANN.ASTRON. 31,43.
51 SPITE,M. 1968,ANN.ASTRON. 31,269. (THESE 1968)
52 GRIFFIN,R. AND GRIFFIN,R. 1967,MONTHLY NOTICES ROY.ASTRON.SOC. 137,253.
53 SPITE,M. 1967,ANN.ASTRON. 30,211.

REFERENCES TO THE CATALOGUE (CONTINUED)

54 SPITE,M. 1969,ASTRON.ASTROPHYS. 1,52.
55 SPITE,M. 1967,ANN.ASTRON. 30,685.
56 HARMER,D.L. AND PAGEL,B.E.J. 1969,NATURE 225,349.
57 CATCHPOLE,A.M.,PAGEL,B.E.J. AND POWELL,A.L.T. 1967,MONTHLY NOTICES ROY.
    ASTRON.SOC. 136,403.
58 COHEN,J.G. AND STROM,S.E. 1968,ASTROPHYS.J. 151,623.
59 BASCHEK,B.,HOLWEGER,H.,NAMBA,O. AND TRAVING,G. 1967,Z.ASTROPHYS.65,418.
60 GRIFFIN,R. 1969,MONTHLY NOTICES ROY.ASTRON.SOC. 143,223.
61 HEARNSHAW,J.B. 1974,ASTRON.ASTROPHYS. 34,263.
62 HEARNSHAW,J.B. 1974,ASTRON.ASTROPHYS. 36,191.
63 HEARNSHAW,J.B. 1973,ASTRON.ASTROPHYS. 29,165.
64 MERCHANT,A. 1966,ASTROPHYS.J. 143,336.
65 STROM,S.E. AND STROM,K.M. 1966,ASTRON.J. 71,181.
66 NISHIMURA,S. 1966,COLLOQUIUM ON LATE-TYPE STARS 125.
67 ADELMAN,S.J. AND SARGENT,W.L.W. 1972,ASTROPHYS.J. 176,671.
68 WARREN,N. AND PEAT,D.W. 1972,ASTRON.ASTROPHYS. 17,450.
69 KODAIRA,K. 1973,ASTRON.ASTROPHYS. 22,273.
70 CONTI,P.S. AND STROM,S.E. 1968,ASTROPHYS.J. 152,483.
71 BUTCHER,H.R. 1972,ASTROPHYS.J. 176,711.
72 KODAIRA,K.,GREENSTEIN,J.L. AND OKE,J.B. 1969,ASTROPHYS.J. 155,525.
73 FARAGGIANA,R. AND VAN"T VEER-MENNERET,C. 1971,ASTRON.ASTROPHYS. 12,258.
74 SPINRAD,H. AND LUEBKE,W.R. 1970,ASTROPHYS.J. 160,1141.
75 POWELL,A.L.T. 1970,MONTHLY NOTICES ROY.ASTRON.SOC. 148,477.
76 SNEDEN,C. 1973,ASTROPHYS.J. 184,839.
77 PATCHETT,B.E.,MCCALL,A. AND STICKLAND,D.J. 1973,MONTHLY NOTICES ROY.ASTRON
    .SOC. 164,329.
78 WILLIAMS,P.M. 1973,MONTHLY NOTICES ROY.ASTRON.SOC. 162,235.
79 WOLFFRAM,W. 1972,ASTRON.ASTROPHYS. 17,17.
80 FRENCH,V.A. AND POWELL,A.L.T. 1970,ROY.OBSERV.BULL.GREENWICH 173.
81 SIEVERS,H.C. 1969,PUBL.ASTRON.SOC.PACIFIC 81,33.
82 BURBIDGE,G.R. AND BURBIDGE,E.K. 1956,ASTROPHYS.J. 124,130.
83 BALL,C. AND PAGEL,B.E.J. 1967,OBSERVATORY 87,19.
84 GRIFFIN,R. 1971,MONTHLY NOTICES ROY.ASTRON.SOC. 155,139.
85 ALEXANDER,J.B. 1967,MONTHLY NOTICES ROY.ASTRON.SOC. 137,41.
86 HEARNSHAW,J.B. 1972,MEM.ROY.ASTRON.SOC. 77,55.
87 HEARNSHAW,J.B. 1971,ASTROPHYS.J. 168,109.
88 PROVOST,J. AND VAN"T VEER-MENNERET,C. 1969,ASTRON.ASTROPHYS.2,218.
89 PAGEL,B.E.J. 1966,COLLOQUIUM ON LATE-TYPE STARS 133.
90 ZIELKE,G. 1970,ASTRON.ASTROPHYS. 6,206.
91 FALIPOU,M.A. 1973,ASTRON.ASTROPHYS. 22,445.
92 HACK,M. 1960,MEM.SOC.IT. 31,279.
93 ALLER,L.H. AND ROSS,J.E. 1970,ASTROPHYS.J. 161,189.
94 ALLER,L.H. AND ROSS,J.E. 1967,MAGNETIC AND RELATED STARS 339.
95 ALLER,M.F. 1972,ASTRON.ASTROPHYS. 19,248.
96 TOMKIN,J. 1972,MONTHLY NOTICES ROY.ASTRON.SOC. 156,349.
97 STROM,S.E.,GINGERICH,O. AND STROM,K.M. 1966,ASTROPHYS.J. 146,880.
98 SPITE,M. AND SPITE,F. 1973,ASTRON.ASTROPHYS. 23,63.
99 BLANC-VAZIAGA,M.J.,CAYREL,G. AND CAYREL,R. 1973,ASTROPHYS.J. 180,871.
100 PRADERIE,F. 1968,ANN.ASTRON. 31,15.
101 KAUFMANN,J.P.,SCHONBERNER,D. AND RAHE,J. 1974,ASTRON.ASTROPHYS. 36,201.
102 CHROMEY,F.R. 1969,ASTROPHYS.J. 158,599.
103 PRZYBYLSKI,A. AND BURNICKI,A. 1974,ACT.ASTRON.24,275.
104 PETERS,G.J. AND ALLER,L.H. 1970,ASTROPHYS.J. 159,525.
105 HARDORP,J. AND SCHOLZ,M. 1970,ASTROPHYS.J.SUPPL. 19,193.
106 KODAIRA,K. AND SCHOLZ,M. 1970,ASTRON.ASTROPHYS. 6,93.
107 GEHLICH,U.K. 1969,ASTRON.ASTROPHYS. 3,169.

REFERENCES TO THE CATALOGUE (CONTINUED)

108 LEE,P. 1974,ASTROPHYS.J. 192,133.
109 WARNER,B. 1966,MONTHLY NOTICES ROY.ASTRON.SOC. 133,389.
110 ADELMAN,S.J. 1973,ASTROPHYS.J. 182,531.
111 HUNDT,E. 1972,ASTRON.ASTROPHYS.21,413.
112 VAN PARADIJS,J. AND DE RUITER,H. 1973,ASTRON.ASTROPHYS.20,169.
113 RODGERS,A.W. 1969,MONTHLY NOTICES ROY.ASTRON.SOC. 145,151.
114 PRZYBYLSKI,A. 1972,MONTHLY NOTICES ROY.ASTRON.SOC. 159,155.
115 PRZYBYLSKI,A. 1970,MONTHLY NOTICES ROY.ASTRON.SOC. 146,71.
116 WOLF,B. 1971,ASTRON.ASTROPHYS. 10,383.
117 STROM,S.E.,GINGERICH,O. AND STROM,K.M. 1968,OBSERVATORY 88,160.
118 ENGIN,S. 1974,ASTRON.ASTROPHYS. 32,93.
119 SCHOLZ,M. 1967,Z.ASTROPHYS. 65,1.
120 REGO,M.E. 1970,URANIA 271,3.
121 FENANDEZ-FIGUEROA,M.J. 1973,URANIA 277,3.
122 LATHAM,D.W. 1970,SMITHSON.INST.ASTROPHYS.OBS.RES.SPACE SCI.REP. 321.
123 SMITH,M.A. 1971,ASTRON.ASTROPHYS. 11,325.
124 SMITH,M.A. 1974,ASTROPHYS.J. 189,101.
125 SMITH,M.A. 1973,ASTROPHYS.J.SUPPL. 25,277.
126 BAKOS,G.A. 1971,J.ROY.ASTRON.SOC.CAN. 65,222.
127 ENGIN,S. 1974,IN PRESS.
128 BELL,R.A. AND BRANCH,D. 1971,MONTHLY NOTICES ROY.ASTRON.SOC. 153,57.
129 PRZYBYLSKI,A. 1968,MONTHLY NOTICES ROY.ASTRON.SOC. 139,313.
130 STROM,S.E.,STROM,K.M. AND SARGENT,W.L.W. 1970,ASTROPHYS.J. 157,1265.
131 WARREN,P.R. 1973,MONTHLY NOTICES ROY.ASTRON.SOC. 161,427.
132 VAN PARADIJS,J. 1973,ASTRON.ASTROPHYS. 23,369.
133 KRIZ,S. 1966,BULL.ASTRON.INST.CZECH. 17,175.
134 ZVERKO,J. 1968,BULL.ASTRON.INST.CZECH. 24,71.
135 HARMER,D.L.,PAGEL,B.E.J. AND POWELL,A.L.T. 1970,MONTHLY NOTICES ROY.ASTRON
     .SOC. 150,409.
136 KIPPER,T. 1969,PUBL.TARTU ASTR.OBS. 36,227.
137 ENGIN,S. 1974,ASTROPHYS.SPACE SCI. 29,343.
138 STROM,S.E.,STROM,K.M. AND CARBON,D.F. 1971,ASTRON.ASTROPHYS. 12,177.
139 GLEBOCKI,R. 1972,ACT.ASTRON. 22,141.
140 WILLIAMS,P.M. 1974,MONTHLY NOTICES ROY.ASTRON.SOC. 167,359.
141 WALLERSTEIN,G.,STONE,Y.H. AND WILLIAMS,J.A. 1962,ASTROPHYS.J. 135,459.
142 ADELMAN,S.J. 1973,ASTROPHYS.J. 183,95.
143 CONTI,P.S. AND LOONEN,J.P. 1970,ASTRON.ASTROPHYS. 8,197.
144 AUER,L.H. 1964,ASTROPHYS.J. 139,1148.
145 CONTI,P.S. AND STROM,S.E. 1968,ASTROPHYS.J. 154,975.
146 ZIMMERMANN,R.E.,ALLER,L.H. AND ROSS,J.E. 1971,ASTROPHYS.J. 161,179.
147 KODAIRA,K.,GREENSTEIN,J.L. AND OKE,J.B. 1970,ASTROPHYS.J. 159,485.
148 SELIGMAN,C.E. AND ALLER,L.H. 1970,ASTROPHYS.SPACE SCI. 9,461.
149 HACK,M. 1969,ASTROPHYS. SPACE SCI. 5,403.
150 PERRIN,M.N.,CAYREL,R. AND CAYREL DE STROBEL,G. 1975,ASTRON.ASTROPHYS.
     39,97.
151 OINAS,V. 1974,ASTROPHYS.J.SUPPL. 27,391.
152 FOY,R. 1974,THESIS.
153 BELL,R.A. AND RODGERS,A.W. 1965,MONTHLY NOTICES ROY.ASTRON.SOC. 129,127.
154 STICKLAND,D.J. 1973,MONTHLY NOTICES ROY.ASTRON.SOC. 161,193.
155 HUNGER,K. 1960,Z.ASTROPHYS. 49,129.
156 GREENSTEIN,J.L. 1948,ASTROPHYS.J. 107,151.
157 TOMLEY,L.J.,WALLERSTEIN,G. AND WOLFF,S.C. 1970,ASTRON.ASTROPHYS. 9,380.
158 DUFTON,P.L. 1973,ASTRON.ASTROPHYS. 28,267.
159 KOHL,K. 1964,Z.ASTROPHYS. 60,115.
160 STROM,K.M. 1969,ASTRON.ASTROPHYS. 2,182.
161 HARDORP,J. 1966,Z.ASTROPHYS. 63,137.

REFERENCES TO THE CATALOGUE (CONTINUED)

162 HARDORP,J.,BIDELMAN,W.P. AND PROLSS,J. 1968,Z.ASTROPHYS. 69,429.
163 HENSBERGE,H. AND DE LOORE,C. 1974,ASTRON.ASTROPHYS. 37,367.
164 GRIFFIN,R. 1975,MONTHLY NOTICES ROY.ASTRON.SOC. 171,181.
165 PASINETTI-FRACASSINI,L. 1975 IN PREPARATION
166 ZVERKO,J. 1970,BULL.ASTRON.INST.CZECH.21,56.
167 WRIGHT,K.O. 1951,PUBL.DOM.ASTROPHYS.OBS.VICTORIA 8,1.
168 YAMASHITA,Y. 1965,PUBL.ASTRON.SOC.JAPAN 17,55.
169 VAN'T VEER-MENNERET,C. 1963,ANN.ASTRON. 26,289.
170 DANZIGER,I.J. 1966,ASTROPHYS.J. 143,591.
171 BUSCOMBE,W.,CHAMBLISS,C.R. AND KENNEDY,P.M. 1968,MONTLY NOTICES ROY.
    ASTRON.SOC. 140,369.
172 AYDIN,C. 1972,ASTRON.ASTROPHYS. 19,369.
173 VAN PARADIJS,J. AND MEURS,E.J.A. 1974,ASTRON.ASTROPHYS. 35,225.
174 BASCHEK,B. AND SEARLE,L. 1970,ASTROPHYS.J. 155,537.
175 KONDO,Y. AND MAC CLUSKEY,G.E. 1969,ASTROPHYS.J. 156,1007.
176 TOY,L.G.S. 1969,ASTROPHYS.J. 158,1099.
177 HYLAND,A.R. AND MOULD,J.R. 1974,ASTROPHYS.J. 187,277.
178 GREENE,T.F.,PERRY,J.,SNOW,T.P. AND WALLERSTEIN,G. 1973,ASTRON.ASTROPHYS.
    22,293.
179 GUTHRIE,B.N.G. 1966,ROY. OBSERV.EDINBURGH 5,181.
180 GUTHRIE,B.N.G. 1967,ROY.OBSERV.EDINBURGH 6,1.
181 MACKLE,R.,HOLWEGER,H.,GRIFFIN,R. AND GRIFFIN,R. 1975,ASTRON.ASTROPHYS.
    38,239.
182 HILL,P.W. 1965,MONTHLY NOTICES ROY.ASTRON.SOC. 129,137.
183 ALLER,L.H. AND BIDELMAN,W.P. 1964,ASTROPHYS.J. 139,171.
184 BACJAR,R. 1969,CONTR.ASTRON.OBS.SKALNATE PLESO 4,63.
185 WOLF,B. 1972,ASTRON.ASTROPHYS. 20,275.
186 WILLIAMS,P.M. 1972,MONTHLY NOTICES ROY.ASTRON.SOC. 155,17P.
187 SNIJDERS,M.A. 1969,ASTRON.ASTROPHYS. 1,452.
188 COWLEY,C.R. 1968,ASTROPHYS.J. 153,169.
189 GROTH,H.G. 1961,Z.ASTROPHYS. 51,206.
190 SEARLE,L.,LUNGERSHAUSEN,W. AND SARGENT,W. 1966,ASTROPHYS.J. 145,141.
191 DA SILVA,L. 1975,ASTRON.ASTROPHYS.41,287.
192 JUGAKU,J.,SARGENT,W.L.W. AND GREENSTEIN,J.L. 1961,ASTROPHYS.J. 134,781.
193 SPITE,F. AND SPITE,M. 1975,ASTRON.ASTROPHYS. 40,141.
194 SELVELLI,P.L. 1972,ASTRON.ASTROPHYS. 20,325.
195 KIPPER,T. 1969,PUBL.TARTU ASTR.OBS.TEATED 21.
196 SCHONBERNER,D. AND WOLF,R.E.A. 1974,ASTRON.ASTROPHYS. 37,87.
197 CONTI,P.S. 1970,ASTRON.ASTROPHYS. 7,213.
198 LUUD,L. AND KUUSK,T. 1970,PUBL.TARTU ASTR.OBS. 38,115.
199 HUNGER,K. AND KLINGLESMITH,D. 1969,ASTROPHYS.J. 157,721.
200 WEGNER,G. AND PETFORD,A.D. 1974,MONTHLY NOTICES ROY.ASTRON.SOC. 168,557.
201 PRZYBYLSKI,A. 1971,MONTHLY NOTICES ROY.ASTRON.SOC. 153,111.
202 ALLER,M.H. 1970,ASTRON.ASTROPHYS. 6,67.
203 WOLF,R.E.A. 1973,ASTRON.ASTROPHYS. 26,127.
204 HUNGER,K. AND KAUFMANN,J.P. 1973,ASTRON.ASTROPHYS. 25,261.
205 VILHU,O.   ,ANN.ACAD.SCI.FENNICAE SERIE A VI PHY. 394.
206 SCHMITT,   1973,ASTRON.ASTROPHYS.SUPPL. 9,427.
207 WOLF,B. 1973,ASTRON.ASTROPHYS. 28,335.
208 HEARNSHAW,J.B. 1975,ASTRON.ASTROPHYS. 38,271.
209 MONTGOMERY,E.F. AND ALLER,L.H. 1969,PROC. NAT.ACAD. 63,1039.
210 GARCIA,Z.L. 1968,Z.ASTROPHYS. 68,278.
211 KHOKLOVA,V.L.,ALIYEV,S. AND RUDENKO,V.M. 1969,IZV.KRYMSK.ASTROFIZ.OBS.
    40,65.
212 HACK,M. 1964,IAU SYMP. 26,227.
213 PRZYBYLSKI,A. 1971,MONTHLY NOTICES ROY.ASTRON.SOC. 152,197.

REFERENCES TO THE CATALOGUE (CONTINUED)

214 AUER,L.H.,MIHALAS,D.,ALLER,L.H. AND ROSS,J.E. 1966,ASTROPHYS.J. 145,153.
215 THOMAS,M. 1971,THESIS.
216 STROHBACH,P. 1970,ASTRON.ASTROPHYS. 6,385.
217 PETERSON,R.C. 1975,ASTROPHYS.J. PREPRINT.
218 GUNN,J.E. AND KRAFT,R. 1962,ASTROPHYS.J. 137,301.
219 LITTLE,S.J. 1974,ASTROPHYS.J. 193,639.
220 CAYREL,G.,CAYREL,R. AND FOY,R. 1975,PREPRINT.
221 PERRIN,M.N. 1975,PREPRINT.
222 CAYREL DE STROBEL,G. UNPUBLISHED.
223 CONTI,P.S.,WALLERSTEIN,G. AND WING,R.F. 1965,ASTROPHYS.J. 142,999.
224 CAYREL DE STROBEL,G.,CHAUVE-GODARD,J.,HERNANDEZ,G. AND VAZIAGA,M. 1970,
    ASTRON.ASTROPHYS. 7,408.
225 ISHIKAWA,M. 1975,PUBL.ASTRON.SOC.JAPAN 27,1.
226 DICKENS,R.J.,FRENCH,V.A.,OWST,P.W.,PENNY,A.J. AND POWELL,A.L.T. 1971,
    MONTHLY NOTICES ROY.ASTRON.SOC. 153,1.
227 REIMERS,D. 1969,ASTRON.ASTROPHYS. 3,94.
228 KOSLOVA,K.I. 1968,ASTROFIZ.ISSLED.IZV.SPE.ASTROFI.OBS. 4,69.
229 PRESTON,G.W. 1961,ASTROPHYS.J. 134,797.
230 MICZAIKA,G.R.,FRANKLIN,F.A.,DEUTSCH,A.J. AND GREENSTEIN,J.L. 1956,
    ASTROPHYS.J. 124,134.
231 BURKHART,C. AND VAN'T VEER,C. 1974,COMPTES RENDUS ACAD.SCI.PARIS SERIE B
    278,1108.
232 BAIRD,S.R.,ROBERTS,W.J.,SNOW,T.P. AND WALLER,W. 1975,PUBL.ASTRON.SOC.
    PACIFIC 87,385.
233 BRANCH,D. AND BELL,R.A. 1970,MONTHLY NOTICES ROY.ASTRON.SOC. 151,289.
234 BELL,R.A. 1972,MONTHLY NOTICES ROY.ASTRON.SOC. 157,147.
235 CONTI,P.S. 1965,ASTROPHYS.J.SUPPL. 11,47.

INDEX OF NAMES

Numbers in bold type refer to invited lectures, italic to contributed papers and ordinary print to the discussion and elsewhere.

Ables, H. D. 196
Abt, H. A. 58, 120, 219, 220
Adam, M. G. XVI
Adams, T. F. 8, 67, 99
Aizenman, M. L. 97
Alexander, J. B. 56
Allen, M. S. 7, 51
Aller, L. H. 198, 199
Andersen, J. 109, 203
Andrillat, Y. *155–156,* 197
Arnett, W. D. 199
Auer, L. H. 49
Auman, J. R. 15
Ayres, T. R. 57
Azzopardi, M. 149

Baglin, A. 159
Barry, D. 58, 191, 192
Baschek, B. **3–15,** 8, 14, 15, 45, 80, 116, 117, 119, 125, 136
Becker, W. XVII, 184
Bell, R. A. 14, 17, 18, 23, **49–65,** 85, 99, 109, 152, 167, 183, 203, 208, 212
Bentolila, C. **223–259**
Bidelman, W. 15, 19, 22, 44, 45, 90, 109, 116, 119, 120, 123, 124, 125, 127, 151, 202, *205,* 213
Blackwell, D. E. 30
Blanc-Vaziaga, M. J. 59, 187, 188, 189, 210
Blanco, C. 163
Bode, G. 5
Boeshaar, G. 203
Böhm, K.-H. 8
Böhm-Vitense, E. 8, 9, 31
Bolton, C. T. *17–18*
Bond, H. E. 57, 121, 123, 124, 125, 193
Boyle, R. J. 192
Branch, D. R. 49, 55, 56, 61, 64
Brosche, P. 168
Bues, I. *23–24*
Burbidge, E. M. and G. R. 198
Burkhart, C. *157–159*
Butler, D. 55, 184, 193, 194, 195, 202

Caloi, V. 97
Cannon, A. 220
Canterna, R. 184, 190, 195, 196
Carbon, D. F. 8, 31, 55, 59, 138
Canavaggia, R. *95–96*
Cayrel, R. 41, 50, 59, 67, 71, 188, 209, 212
Cayrel de Strobel, G. **223–259,** 50, 59, 71, 78, 96, 125, 146, 151, 173, 174, 188, 202, 211, 212, *223–259*
Chalonge, D. *143–146,* 225
Christensen, C. 196
Claria, J. J. *73, 101–109*
Clegg, R. E. S. 49, 55, 61
Code, A. D. XVII
Conti, P. S. 67, 119, 159, 212
Couteau, P. 168
Cowley, C. R. 51, 118, 119
Crawford, D. L. *71–72,* 82
Cromwell, R. 58, 191, 192
Cucchiaro, A. *177–180*

Dachs, I. 151
Da Silva, L. 23, 168, *215–218*
Day, R. W. 57
Dearborn, D. S. 57, 63, 190
Delbouille, L. 60, 63
Delcroix, A. 165
Demarque, P. 97
Deutsch, A. 118, 212
Dickens, R. J. 50, 55, 61, 63
Divan, L. *143–146*
Dubois, P. *149–152*
Dufour, R. 199

Eggen, O. J. XVI, 23, 69, 115, 125, 167, 168, 173, 174, 175, 192, 193, 196, 205, 213
Eriksson, W. C. 50, 59

Feast, M. N. 150, 151
Florsch, A. 149
Forrester, W. T. 197
Foy, R. 14, 50, 63, 64, 125, 167, 168, 173, *209–212*

Fry, M. A. 197
Frye, R. L. 147
Furenlid, K. I. 64, 202

Garrison, R. F. 14, *17–18*, 45, 82, 85, 116, 127, 153, 180
Gascoigne, S. C. B. 197
Gerbaldi, M. 68, 80, 180, 208
Gingerich, O. 31, 138
Glebocki, R. 89
Golay, M. 122
Gordon, K. C. 135, 136
Graham, J. A. 115, 121
Greenstein, J. L. 23, 31, 42, 115, 117, 120, 205
Grenier, S. *215–218*
Grenon, M. *75–78*, 173, 190
Griffin, R. *19–20*, 54, 57, 58, 63, 64, 191, 192, 202, *213*
Gros, M. 179
Gross, P. G. 101, 105, 106
Groth, E. J. 8, 152
Gustafsson, B. 8, 9, 31, 50, 52, 54, 56, 59, 61, 64, 184, 190, 202, 211
Gyldenkerne, K. 42

Hack, M. 45, 63, 125, 127
Hagen, J. P. 127
Hall, D. N. R. 57
Hansen, L. 56, 173, **207**
Harmer, D. L. 56, 190
Hartoog, M. R. 51
Hartwick, F. D. A. 60, 195, 196
Hauck, B. *67–69*, 80, 164, *223–259*
Hauge, O. 51
Hearn, A. 14
Hearnshaw, J. 34, 49, 53, 55, 185, 190
Heasley, J. N. 97
Heck, A. *215–218*
Hejlesen, P. M. 39, 40, 42, 43, 97, 171
Helfer, H. L. 34, 192, 193
Henry, R. C. 158
Hesser, J. E. 85, 158
Holweger, H. *19–20*, 50, 51
Honeycutt, R. K. 118, 119
Houk, N. 120, 122, 123, *127–128*, 220
Houtgast, J. 57
Houziaux, L. 17, 127, 153, *155–156*, 165
Hoyle, F. 203
Humphries, C. M. *161–165*
Hunger, K. 7, 116
Hutching, J. B. 150
Hyland, A. R. 138, 139

Iriarte, B. 184

Irvine, N. 127
Irwin, A. W. *17–18*

Janes, K. A. 102, 105, 186, 187, 189, 190
Jaschek, C. 15, 45, 46, **113–125**, *149–152*, *177–180*, 202, 224
Jaschek, M. 123, *149–152*, *177–80*, 224
Jenner, D. C. 199
Johnson, H. L. VII, 88, 90, 184, 224
Joly, M. 197
Jones, D. H. P. 194
Jung, J. 216

Kandel, R. 14, 15, 25, 46, 165
Keenan, P. C. XVI, 19, 45, 80, 121, 122, 123, 127, *147*, 203, 219, 220
Keller, G. 121
Kellman, E. XVI
Kinman, T. D. 192, 196
Kjaergaard, P. 56, 61, 173, 203, 207
Knuckles, C. F. 184
Kohlschütter, E. XVI
Kondo, Y. 34
Kraft, R. P. 55, 184, 193, 195, 196
Krishna Swamy, K. S. 6
Kron, G. E. 135, 136
Krupp, B. M. 57, 63
Kuchowicz, B. 119
Kurucz, R. L. 8, 31, 61, 138

Labs, D. 8
Lambert, D. L. 57, 63, 64
Larson, R. B. 196
Leckrone, D. S. 12
Leighton, R. B. 91, 224
Lindblad, B. 192
Luebke, W. R. 190
Luyten, W. 205
Lynden-Bell, D. 193, 196

Macau-Hercot, D. *177–180*
MacConnell, D. J. 123, 124, 147
Mäckle, R. *19–20*, 31, 174
Maeder, A. 46, 63, *97–99*, 172, 208
Magnenat, P. 67, 68
Malaise, D. 178
Malaroda, S. 119
Martinet, L. 211
Maury, A. C. 220
Mayall, N. U. 192
Mayor, M. *207–208*, 212, 216
McCarthy, M. F. 17, 63, 152, 156, 180, 203, 205
McClure, R. D. 60, 102, 107, 108, 186, 187, 190, 192, 193, 195, 196, 197

McCuskey, S. W. 118, 119
Mendoza V., E. E. *79–80*, 93, 127, 172
Meurs, E. J. A. 213
Mianes, P. *95–96*
Mihalas, D. 3, 5, 7, 8, 11, 31, 49
Minnaert, M. H. 57
Mitchell, R. 184, 224
Moore, C. C. 57
Morel, M. *223–259*
Morgan, W. W. XV–XIX, 45, 58, 63, 72, 83, 85, 87, 88, 90, 113, 118, 119, 120, 121, 122, 125, 127, 133, 149, 156, 192, 203, *219–220*
Morton, D. C. 8, 11, 67, 99
Müller, E. A. 44, 51, 64
Münch, G. XVI, 117

Nandy, K. *161–165*
Nariai, K. 6, 118
Nassau, J. J. XVI, 121
Neugebauer, G. 91, 224
Newell, E. B. 101
Nicholson, W. 92
Nikolov, N. 95
Nissen, P. E. 8, 19, 49, 54, 63, 72, *81–83*, 85, 199, 211
Nordlund, A. 50, 59, 61
Noyes, R. W. 57

Ohlmacher, J. 61
Oinas, V. 50, 167, 185, 188, 190
Oke, J. B. 67
Osawa, K. 69, 118
Osborn, W. 19, 55, 60, 65, 69, *73*, *101–109*, 127, 195, 211, 213
Osmer, P. S. 150, 152, 197
Osterbrock, D. XVII

Paczynski, B. J. 57
Pagel, B. 55, 56, 59, 184, 185, 192, 211
Pannekoek, A. 87
Parsons, S. 152
Pasinetti, L. E. *173–175*
Patchett, B. 184, 185
Payne-Gaposchkin, C. *91–93*, 203
Pecker, J. C. 3
Peimbert, M. 198, 199, 202
Penston, M. 185
Perrin, M.-N. 50, *167–172*
Perry, C. L. 61, *71–72*
Pesch, P. 22
Peterson, R. 188, 189, 202
Petford, A. D. 135, 140
Pettit, E. 92
Peytremann, E. 8, 9, 31, 49, 54, 61, 63, 177, 211

Philip, A. G. D. 101
Praderie, F. 159
Prather, M. J. 97
Preston, G. W. 120, 193
Przybylski, A. 56, *135–141*, 151, 152, 197

Querci, F. 14, 25, 64

Racine, R. 97, 102
Raff, M. 190
Redman, F. 58, 184
Reimers, D. 8
Reza, de la, R. 14, 15, 80
Robinson, L. 193
Roman, N. G. XVI, 88, 90, 120
Rood, R. T. 60
Rosendahl, J. D. 150
Rousseau, J. *95–96*
Rudkjøbing, M. 127, 159

Sandage, A. R. 69, 96, 192, 193, 196
Sanduleak, N. 73, 149, 203
Sargent, W. L. W. 115, 116, 117
Sauval, A. J. *21–22*
Schatzman, E. 14, 85
Scheid, P. S. 216
Schild, R. 67
Schlesinger, B. M. 97
Schmidt, E. G. 95, 96, 122, 162
Scholz, M. 13
Schönberner, D. 8
Schwarzschild, M. and B. 192
Schweizer, F. 196
Searle, L. 115, 117, 119, 198
Seitter, W. 64, 115, *129–133*
Sharpless, S. XVII
Shipman, H. L. 23
Slettebak, A. 115, 119, 121
Smethells, W. G. *205*
Smith, H. E. 197, 198, 202
Smith, M. 159
Sneden, C. 55, 57, 193
Spinrad, H. 20, 45, 49, 57, 58, 80, 83, 96, 99, 141, **183–204**, 210, 213
Spite, F. 34, 43, 215
Spite, M. 43, 216, 218
Steinlin, U. 43, 184, 202
Stephenson, C. B. 17, 73, 83, 203
Stock, J. 73
Stokes, N. R. 57
Stone, R. P. S. 197, 198
Strittmatter, P. A. 23
Strohbach, P. 167
Strom, S. E. 8, 59, 138, 189

Strömgren, B. **XV—XVIII,** 42, 61, 71, 81, 219
Sturch, C. 193
Swings, J. P. 177, 179

Talbot, R. J. 199
Tammann, G. A. 96
Taylor, B. J. 49, 57, 184, 189, 190, 191, 197,
   210
Thompson, G. I. *161–165*
Thomsen, B. 83
Tinsley, B. 211
Titus, J. XVI, 119
Tomkin, J. 57, 64
Torres-Peimbert, S. 97, 199
Traving, G. 7, 13
Tsarevsky, G. S. 95
Tsuji, T. 21, 211

Unsöld, A. 3, 11, 143
Upgren, A. R. 125, 147
Upson, W. L. 57

Van den Bergh, S. 195, 196, 197
Van Paradijs, J. 213
Van 't Veer, Cl. 68, *157–159*
Vauclair, S. 85, 159
Veth, C. 18

Walborn, N. 19, 44, 45, 83, *85*, 116, 125, 133,
   152, *153*, 202
Wallerstein, G. 34, 192
Wampler, E. J. 193
Warner, B. 34, 123, 198
Wegner, G. 23, 135, 140
Wehrse, R. 13, *25*
Welch, G. 197
Westerlund, B. 96
Whitford, A. E. XVII
Whitney, C. A. 91
Wickramasinghe, N. C. 23
Wildey, R. L. 138
Williams, P. M. 56, 62, 64, 72, *87–90*, 109,
   125, 127, 147, 173, 175, 183, 184, 189, 190,
   202, 211, 213
Willstrop, R. V. 56
Wilson, O. C. 121, 122, 147
Wing, R. F. 57
Wolf, B. 8, 151, 152
Woodrow, J. E. J. 15
Woolley, R. 38
Wroblewski, H. 73

Yakimova, N. N. 95

Zinn, R. 55, 196